T0314048

Cyber-Physical Distributed Systems

Cyber-Physical Distributed Systems

Modeling, Reliability Analysis and Applications

Huadong Mo
School of Engineering and Information Technology
University of New South Wales
Canberra, Australia

Giovanni Sansavini
Institute of Energy and Process Engineering
Department of Mechanical and Process Engineering
ETH Zurich
Zurich, Switzerland

Min Xie
Department of Systems Engineering and Engineering Management
City University of Hong Kong
Hong Kong, China

Registered Offices
John Wiley & Sons, Inc., 111 River Street, Hoboken, NJ 07030, USA
John Wiley & Sons Ltd, The Atrium, Southern Gate, Chichester, West Sussex, PO19 8SQ, UK

Editorial Office
The Atrium, Southern Gate, Chichester, West Sussex, PO19 8SQ, UK

For details of our global editorial offices, customer services, and more information about Wiley products visit us at www.wiley.com.

Wiley also publishes its books in a variety of electronic formats and by print-on-demand. Some content that appears in standard print versions of this book may not be available in other formats.

Library of Congress Cataloging-in-Publication Data applied for
Hardback ISBN: 9781119682677

Cover Design: Wiley
Cover Image: © Wright Studio/Shutterstock

Set in 9.5/12.5pt STIXTwoText by Straive, Chennai, India
Printed and bound by CPI Group (UK) Ltd, Croydon, CR0 4YY

C9781119682677_020821

Contents

Preface

A cyber-physical system (CPS) consists of a collection of computing devices communicating with one another and interacting with the physical world via sensors and actuators in a feedback loop. Increasingly, such systems are everywhere, from smart buildings to medical devices to automobiles. The emergence of CPSs as a novel paradigm has revolutionized the relationship between humans, computers, and the physical environment. CPSs are still in their infancy, and most recent studies are application-specific and lack systematic design methodology. As a result, it is challenging to investigate and explore the core system science perspective needed to design and build complex CPSs, which are of great importance in many applications.

Using the underlying theories of systems science, such as probability theory, decision theory, game theory, control theory, data analysis, organizational sociology, behavioral economics, and cognitive psychology, this book addresses foundational issues central across CPS applications, including: (I) System Verification – How to develop effective metrics and methods to verify and certify large and complex CPSs; (II) System Design – How to design CPSs to be safe, secure, and resilient in rapidly evolving environments; (III) Real-Time Control and Adaptation – How to achieve real-time dynamic control and behavior adaptation in diverse environments, such as distribution and in network-challenged spaces; (IV) System of Systems – How to harness communication, computation, and control for developing new integrated systems, reducing concepts to realizable designs, and producing integrated software–hardware systems at a pace far exceeding today's timeline.

In general, this book has four essential topics. Chapters 1 and 2 provide readers who do not have a sufficient background on CPSs with a general introduction, research gaps, and representative CPS applications, including CPS modeling, statistical analysis of CPS performance, probability prediction of CPS state, robust CPS control techniques, and management and optimization of CPS reliability and risk. Chapters 3 and 4 mainly concern the robust control of CPSs by designing optimal control strategies, or resource management to enhance robust performance and improve the reliability index against time delays and packet dropouts, which are the inherent properties of open communication networks. Chapter 5 addresses the data-driven degradation modeling of aging physical (actuators) and cyber (sensors) components of CPSs, and corresponding optimal maintenance plans to improve the reliability of CPSs. Chapters 6 and 7 investigate the cyber security of CPSs, introduce the general concept of cyberattacks, design vulnerability models, and risk assessment procedures, and develop game-theoretic mitigation techniques and Bayesian-based cyberteam deployment strategies.

More specifically, Chapter 1 summarizes the evolution from the traditional physical system to the CPS and provides an overview of dynamic and dependent behaviors to be addressed in the subsequent chapters of the book. The introduction discusses some important and recent challenges in improving traditional physical systems in terms of CPSs, popular research trends in evaluating the impacts of CPSs on society, and opportunities for enhancing the performance of realistic applications, which are primarily network control systems. The detailed properties, requirements, and vulnerabilities of utility systems are also introduced. The reasons why the proposed modeling techniques work is important in a field that would be difficult to deal with if the cyber and physical domains were treated separately.

In Chapter 2, readers acquire the basic knowledge to be used in data-driven statistical modeling, the estimation of the probabilistic CPS state, and a comprehensive framework for conducting reliability analysis of CPSs. In addition, this chapter introduces how to use to historical data to validate the performance of the proposed CPS model, and how to use performance indexes to facilitate the resilient design of CPSs. Moreover, it also demonstrates a real-time test platform for various industrial applications and the standard procedures for improving real-time criteria.

Chapter 3 focuses on the stability of CPSs, where decision makers perform dynamic control and adaptation based on real-time data from sensors. It provides two examples of the design of the controller parameters for robust system performance. The first example illustrates the development of adaptive control for wide-area measurement power systems, where communication delays are predicted to provide delay compensation for additional frequency stability. In addition, the integration of control theory, power engineering, and statistical estimation is discussed. The second example is an extension of wide-area measurement power systems from a dedicated communication network to open communication networks, where occurrences of communication delays and packet dropouts result in the failure of the power management system from renewable energy resources. Explicit and implicit methods are then designed for system integration, analysis, and improvement.

Chapter 4 illustrates a system-of-systems framework for the reliability of distributed CPS accounting for the impact of degraded communication networks. This is quite different from the focus of Chapter 3, which mainly covers the stability of CPSs from a control perspective. Based on the collected dataset, the degradation path of open communication networks is described in terms of stochastic continuous time transmission delays and packet dropouts. A distributed generation system with open communication infrastructure is used as an example, which is a multi-area distributed system that is more complicated than the single-area power system presented in Chapter 3. An optimal power flow model is proposed to generate consecutive time-dependent optimal operation scenarios for a distributed CPS. Quantitative analysis is carried out to evaluate the effect of networked degradation on the reliability indexes of CPSs, e.g., energy not supplied and operation cost. A prediction method for reconstructing missing data is proposed to mitigate the influence of packet dropouts, which is universal and applicable to most current industrial applications.

Chapter 5 models the functional dependence between stochastic aging actuators and sensors within their operating environments. This dependence is considered in the time domain, causing a distinct degradation status in the actuators and sensors. Reliability modeling of the stochastic effects and effective maintenance activities are discussed for different types of CPSs, including the cooling system in a nuclear power plant, a one-area energy system with a single generation group, and a multi-area energy system with several different generation groups.

Chapter 6 explores the concepts, principles, practices, components, technologies, and tools behind risk management for cybersecurity of CPSs, providing practical experience through a realistic case study that focuses on the methodologies available to identify and assess such threats, evaluate their impact, and determine appropriate measures to prevent, mitigate, and recover from any threat or disruptive event so that the operations and profitability of the organizations are maintained and maximized.

Chapter 6 presents the framework of CPSs under cyberattacks from a game-theoretic perspective, which makes use of statistical data to model the behavior of cyberattacks and study the dynamic game between the network defender and attacker at the system level. For current utility CPSs, cyber threats from supervisory control and data acquisition (SCADA) systems, and spear-phishing attacks on the accounts of internal employees to gain access to dedicated communication networks are investigated in Chapters 6 and 7. In addition, these chapters focus on how to modify the basic modeling techniques presented in Chapter 2 to describe cyber vulnerabilities, such as seizing the SCADA system under control, disabling/destroying IT infrastructure components, and denial-of-service (DoS) attacks on the control center in smart grids. Based on historical data for IT security spending, the cost of launching distributed DoS attacks, and the occurrence probability of cyber event losses, the contest intensity between the attacker and defender can be accurately predicted for the next period to guide the design of an effective network protection plan, that is, a game-theoretic protection plan and a Bayesian-based cyberteam deployment.

Chapter 7 investigates sequential control problems (i.e., sequential cyberteam deployment) in modern CPSs by introducing an adversarial cost sequence with a variation constraint. Chapter 6 reviews the data-driven vulnerability model, and Chapter 7 deals with the dataset of the arrival time of cyberattacks, which uncovers the statistical pattern of attackers. To solve such problems, a fundamental idea is to first obtain sampled parameters for the arrival model of cyberattacks from the posterior distribution of realistic cyberattack arrival records. The reinforcement learning model for estimating parameters is formulated as a partly parameterized Bayesian model. As a result, the sampled parameters are used instead of the true parameters. The paradigm of this framework can also be applied to other classical models, although specific models are used here for illustration purposes. Next, a Bayesian multi-node bandit is built to cope with the problem, and an online learning algorithm (the Thompson-Hedge algorithm) is forwarded to retain a converging regret function that is a function of the cyberteam deployment. By comparison with the existing algorithm, the convergence rate of the regret function in the proposed algorithm is found to be superior.

Each of Chapters 3–7 can be read independently when one is interested in a specific type of application or further research, making the book attractive to readers from different areas or positions. This book has the following distinct features:

- It is the first book to systematically focus on CPSs with respect to modeling and reliability analysis.
- It provides a comprehensive treatment of imperfect fault coverage (single level/multi-level or modular), functional dependence, common cause failures (deterministic and probabilistic), competing failures (deterministic and probabilistic), and dynamic standby sparing.

- It includes abundant illustrative examples and case studies based on real-world systems.
- It covers recent advances in combinatorial models and algorithms for CPS modeling and analysis.
- It has a rich set of references, providing helpful resources for readers to pursue further research and study of the topics.

The target audience of the book is undergraduate (senior level) and graduate students, engineers, and researchers in system science and related disciplines, including those in computers, telecommunications, transportation, and other industries. Readers should have some knowledge of basic probability theory, control theory, computer science, optimization, game theory, and stochastic processes. However, the book includes a chapter reviewing the fundamentals that readers need to know for understanding the content of the other chapters covering advanced topics in CPSs and case studies. The book can provide readers with knowledge and insight of CPS behaviors, as well as the skills of modeling and analyzing these behaviors to guide the resilient design of real-world critical systems. Thus, the book includes necessary background information, making it self-contained. For some detailed topics, selected for their importance and application potential, references are provided for those interested in further details.

We would like to express our sincere appreciation to the many researchers who have proposed some underlying concepts, frameworks, and methods used in this book, or who have co-authored with us some topics of the book and provided their insights; to name a few, Professor Enrico Zio from Politecnico di Milano, Professor Yanfu Li from Tsinghua University, Dr. Gregory Levitin from the Israel Electric Corporation, Professor David W. Coit from Rutgers University, and Professor Junlin Xiong from the University of Science and Technology of China. Though there are many other researchers to mention, and we have tried to recognize their significant contributions in the bibliographical references of the book.

Finally, it was our huge pleasure to work with Juliet Booker, managing editor of Electrical & Computer Engineering, John Wiley & Sons Ltd., and her team, who have assisted in the publication of this book. We deeply appreciate their efforts and support.

December 15, 2020

Huadong Mo
Min Xie
Giovanni Sansavini

Acronyms and Abbreviations

ACE	Area Control Error
AERs	All-Electric Ranges
AGC	Automatic Generation Control
AGC'	Actual Global Cost
B	Biomass Power
BESS	Battery Energy Storage System
CBM	Condition-Based Maintenance
CC	Control Center
CCC	Centralized Control Centers
CCF	Cross-Correlation Factor
CPS	Cyber-Physical System
CSMA/AMP	Carrier Sense Multiple Access with Arbitration on Message Priority
CSMA/CD	Carrier Sense Multiple Access with Collision Detection
CTMC	Continuous-Time Markov Chains
DC	Direct Current
DEG	Diesel Engine Generator
DERs	Distributed Energy Resources
DG	Distributed Generation
DGS	Distributed Generation Systems
DHMM	Discrete HMM
DoS	Denial of Service
DR	Demand Response
EENS	Expected ENS
EM	Expectation-Maximization
ENS	Energy Not Supplied
ERP	Expected RP
ESS	Energy Storage System
EV	Electrical Vehicle
EWMA	Exponentially Weighted Moving Average
FDs	Feeders
FERC	Federal Energy Regulatory Commission
FESS	Flywheel Energy Storage System
FFT	Fast Fourier Transform

G	Natural Gas Plant
GA	Genetic Algorithms
GC	Global Cost
GM	Grey Differential Model
Ls	Loads
LAN	Local Area Networks
LC	Long-Run Cost Rate
LFC	Load Frequency Control
LP	Linear Programming
HAN	Home Area Networks
HGSAA	Hybrid Genetic-Simulated-Annealing Algorithm
HMM	Hidden Markov Model
HPS	Hybrid Power System
MAC	Media Access Control
MADT	Maximum Allowable Delay Time
MCR	Maintenance Cost Rate
MCS	Monte Carlo Simulation
MDEM	Missing Data Expectation Maximization
MLE	Maximum Likelihood Estimation
MNB	Multi-Node Bandit
MS	Main Supply
MSE	Mean Squared Errors
MTU	Maximum Transmission Unit
NAN	Neighbourhood Area Networks
NCS	Networked Control System
NHPP	Non-Homogeneous Poisson Process
OPF	Optimal Power Flow
O&M	Operation & Maintenance
PBM	Performance-Based Maintenance
PCLPs	PHEV Charging Load Profiles
PCM	Percentage of Corrective Maintenance
PI	Proportional–Integral
PID	Proportional–Integral–Derivative
PHEVs	Plug-In Hybrid Electric Vehicles
PMU	Phasor Measurement Units
PO	Percentage Overshoot
PPM	Percentage of Preventive Maintenance
PSO	Particle Swarm Optimization
PV	Photovoltaic Power
RBD	Reliability Block Diagram
RERs	Renewable Energy Resources
RP	Redundant Power
RT	Rising Time
RTU	Remote Terminal Unit
RUL	Remaining Useful Lifetime

SA	Simulated Annealing
SCADA	Supervisory Control and Data Acquisition
SOC	State of Charge
ST	Settling Time
TCo	Total O&M Cost
TENS	Total ENS
W	Wind Power
WAMS	Wide-Area Measurement Systems
WAN	Wide Area Networks
WAPS	Wide-Area Power System
WCSS	Within-Cluster Sum of Square
WLAN	Wireless Local Area Networks
WTG	Wind Turbine Generator

1

Introduction

In this chapter, descriptions of traditional physical and cyber systems are provided to identify existing challenges. Current research trends of cyber-physical systems (CPSs) are then illustrated to address these challenges. The major applications of the proposed methods in CPSs are reviewed.

1.1 Challenges of Traditional Physical and Cyber Systems

Over the past three decades, studies have addressed numerous concerns regarding the capability of traditional static modeling methodologies, such as the fault tree method and the event tree method, to adequately and quantitatively analyze the impact of hardware and software interaction on the stochastic behavior of CPSs [1, 2]. During the past decade, the dynamical Markov reliability model was proposed to solve similar problems in CPSs [3]. Control block diagrams were presented for cooling loop systems. The reliability block diagram (RBD) was then established and used to describe the overall reliability status of individual components in a simplified form [4, 5]. However, RBDs are incapable of describing the dynamic maintenance and repairable activities; thus, various dynamic modeling methods have been reviewed in [6, 7]. The Markov methodology has the advantage of tracking the dynamic changes and time-dependent features of CPSs, and simply integrates all failure states that occur after each working state into one failure state. The Markov methodology eliminates most of the failure states into a system failure state (absorbing node) by conducting a necessary fault injection test and achieving a sparse transfer matrix but may still result in a very large model due to many existing surviving states. Its modeling precision largely depends on the number of fault injection tests, and more cycles yield higher accuracy. To avoid the disadvantages of these two methodologies, some studies have proposed hybrid reliability models combining RBDs and Markov models for CPSs [8].

The control block diagram introduces blocks to represent each part of the control system, including the controllers, actuators, and control objectives. Control block diagrams are widely used in modern control systems because they can visually describe the relations among the important components, data flow, and control sign flow. In addition, compared with other mathematical models, they have the advantage of simply reflecting the actual correlations in a CPS. It is reasonable to build a reliability model based on the control block diagram of a CPS. In the model, the controller has many input signals, including commands

Cyber-Physical Distributed Systems: Modeling, Reliability Analysis and Applications, First Edition.
Huadong Mo, Giovanni Sansavini and Min Xie.
© 2021 John Wiley & Sons Ltd. Published 2021 by John Wiley & Sons Ltd.

and system state feedback. In general, commands are the system's expected outputs. Control signal flows are given in the control block diagram, and sensors play an important role in this feedback system. This control block diagram clearly indicates the internal dynamic relations of the system, covering most of the aspects that need to be studied.

For applications in CPSs, we are interested in real-time performance. Therefore, from a control perspective, the ability to adjust the transient and steady-state response of a feedback CPS is a beneficial outcome of the design of the CPS. One of the first steps in the design process is to specify the performance measures. In this chapter, we introduce common time-domain specifications, such as percent overshoot, settling time, time to peak, time to rise, and steady-state tracking error. We will use selected input signals, such as the step and ramp, to test the response of the CPS. The correlations between the system performance and the stability, reliability, and resilience strategies of CPSs are investigated. We will develop valuable relationships between the performance specifications and the component states for CPSs.

The ability of a feedback CPS to compensate for the consequences of the inherent faults redefines the concept of failures, i.e., the reliability of the CPS is dependent not only on the type of failure that may occur, but also on the evolving states of system output and control signals in each period [9, 10]. Classical reliability evaluation methods, such as fault tree analysis, event tree analysis, and failure mode and effect analysis, are not appropriate for application to these evolving states due to the level of complexity and dynamics of CPSs. In [11, 12], structured analyses and design techniques based on Monte Carlo simulation (MCS) for reliability evaluation are presented. This approach explicitly formalizes the functional interactions between subsystems, identifies the characteristic values affecting the reliability of complex CPSs, and quantifies the reliability, availability, maintainability, and safety (RAMS) parameters related to the operational architecture. As the remaining ability of the system to maintain the expected control goal after faults occur is crucial, ordered sequences of multi-failure methods have been applied to assess the reliability of all possible CPS architectures [10]. A new methodology called a multi-fault tree is proposed, and time-ordered sequences of failures are addressed.

In contrast to the aforementioned studies, the reliability of a CPS as a function of the required performance from a control viewpoint is evaluated in [13]. The CPS is regarded as a failure if the dynamic performance does not satisfy all the requirements. Difference equations are introduced to describe the stochastic model of the CPS, explicitly illustrating the influence of the transmission delays and packet dropouts on changing the model parameters. A linear discrete-time dynamic approach for modeling the signal flow in, out, and among all subsystems promotes straightforward calculation of fundamental dynamic aspects, such as times and fault characteristics [14].

MCS has been shown to be a straightforward yet accurate approach for the study of such complex systems [11–13, 15]. The general approach in MCS for reliability assessment is to generate operational requirements that lead to the failure of the entire system. However, this approach requires knowledge of the system requirements-to-failure distribution in advance. In [16], an event-based MCS method was proposed for multi-component systems, in which the failure time for each component is generated and then used to verify the success or failure of the system subject to the required operational time. Because no attempt is made to generate the failure time for the entire system, which requires knowledge of the time-to-failure distribution of the entire system as well as the distribution approximation at the component level, it is quite different from previous methods and can reduce the possible error and computational effort in estimating the system reliability.

In [13], this method was extended to estimate the reliability of CPSs and replaces the constraint on the number of replications used in [16] with two other constraints, namely, a precision interval and a percentage of simulations belonging to this interval. The networked degradations for each channel are generated and are then used to determine the success or failure of the CPS for a given combination of operational requirements. Therefore, the reliability of the CPS is estimated as a tabulated function of the operational requirements. Compared with the results in [16], the results obtained in [13] guarantee the estimated reliability to satisfy a given precision.

1.2 Research Trends of CPSs

1.2.1 Stability of CPSs

In power systems, communication delays always occur in the transmission of frequency measurements from sensors to the control center (S-C channel), and control signals from the control center to the plant side (C-A channel) [17, 18]. In local networks, time delays are usually ignored because the control is mainly applied locally, and communication delays are negligible compared with the subsystem time constants [19]. In recent years, with the rapid development of wide-area measurement systems (WAMSs), a large number of phasor measurement units (PMUs) have been deployed to facilitate the real-time control of wide-area power systems (WAPSs) to improve the load frequency control (LFC) performance [20–22]. However, when data packets are transmitted across a WAMS, communication delays may become significant and cannot be ignored [23–25].

The conventional control of WAPSs is centralized and employs dedicated communication channels over a closed communication network. However, new regulatory guidelines require coordination across multiple hierarchical levels of power systems for more effective market operations and, as a result, open communication infrastructures have been deployed to promote the control of these increasingly complex systems [26, 27]. While open networks have economic, maintenance, and reliability advantages, they are subject to time delays that are inherently stochastic (e.g., multiple delays [20] and probabilistic interval delays [24]), and thus cannot be calculated based on the procedures for dedicated networks. Numerical investigations show that time delays the open communication network have the potential to destabilize a WAPS [23, 28, 29]. For instance, the deregulation of the power industry has pushed many tie lines between control areas to operate close to their maximum capacity. This is especially true for tie lines serving heavy load centers, for example, in southern California [28]. Under these circumstances, operational stresses, such as large time delays, increase the possibility of inter-area oscillation, reducing the effectiveness of control system damping, and potentially leading to loss of system synchronism [30].

Recent studies on the LFC of WAPSs with communication delays have mainly focused on: 1) the influence of time delays on the WAPS and effective approaches for delay compensation [31, 32]; 2) control methods for providing robust performance against delays (e.g., event-triggered control methods [33], wide-area phasor power oscillation damping controllers [34], and robust controllers [29, 35]); 3) exact methods for evaluating the delay margin for stability, that is, the maximal allowable delay (upper bound on the time delays), when WAPSs and open networks are integrated [17, 20, 36]. A WAPS is unstable if the real-time delays exceed the delay margin.

Two classes of methods are available for computing the delay margin in a WAPS for constant and time-varying delays. Frequency-domain direct methods quantify the delay margin based on computing critical eigenvalues for constant delays with a known upper bound, for example, the Schur-Cohn-based method for commensurate delays [37], Rekasius substitution [38], and the elimination of exponential terms in the characteristic equation [36]. Indirect methods can deal with time-varying and constant delays; they are derived from Lyapunov stability theory [17, 23, 24], linear matrix inequality techniques [27, 28, 39], H-infinity robust synthesis [20], and the dual-locus diagram method [40].

These methods assume well-defined time delay models, that is, constant [20], uniformly distributed [25], multiple [17], and probabilistic interval time delays [24], and require prior knowledge of the lower bound, upper bound, and parameters of the delay distribution. However, delays in an open communication network vary with the number of active end-users [41] and media access control (MAC) protocols [42]; therefore, no prior knowledge of time delays is available, and they are intrinsically stochastic. These assumptions hinder the application to real systems and lead to excessively conservative results. To overcome these limitations, the LFC model of WAPSs should account for a real open communication network with various MAC-level protocols. The TrueTime simulator has been used to extract realistic scenarios of networked control systems (NCSs), in which the characteristics of the random delays are unknown and are statistically inferred after collecting sufficient delay observations [43–45]. As a specific application of NCSs, insights obtained from the integration of traditional control systems and open communication networks can foster the integration of open communication networks and WAPSs. However, to the best of the author's knowledge, few studies have analyzed the stabilization of WAPSs via real open communication networks.

During the computation of the control signals, the prediction of the C-A delay in the current period and the S-C time delay in the successive period can greatly improve the controller performance. However, this possibility has not been investigated in previous works on delay margins or robust LFC strategies. Several methods for delay estimation are available, such as the Markovian model approach [46], backpropagation neural network prediction [47], adaptive wide-area power oscillation damping [48], dynamic Markov jump filters [49], hidden Markov models (HMMs) [50, 51], and the exponentially weighted moving average (EWMA) method [52]. Time delays depend on the underlying network state, which is ever-changing and concealed [53, 54]. Indeed, network states cannot be observed directly, but random time delays can be measured using the time-stamp technique [55]. Therefore, the state variables of the open communication network can be estimated using the measured time delays; the transitions among these states can be modeled by a discrete HMM (DHMM). As a result, random delays are observations of the DHMM [56–58].

Traditionally, in HMM-based delay models, random delays are mapped to a discrete observation space through scalar quantization techniques, for example, uniform quantization and k-means clustering quantization [57–60]. Because the evolution of random delays is described by a finite-state Markov chain, the communication network action can be learned via the DHMM [47, 57, 59]. In addition, DHMMs have been used for other applications in power systems, for example, the dynamic detection of transmission line outages [61], the generation of distributed photovoltaic systems [62], and residential energy use [63–65]. Chapter 3 introduces the missing data expectation maximization (MDEM)-based Baum-Welch algorithm [56, 59, 66] for the estimate of the parameters of the DHMM online, without previous knowledge of the state transition matrix. Following this step, the time delay can be predicted using the Viterbi algorithm [67]. The predictions are used as the inputs of the Smith predictor to compensate for time delays and improve the LFC strategy performance [68, 69]. Our work improves the conventional Smith predictor, which simply employs the summation of the latest measured C-A time delay and the S-C time delay to construct the prediction [41]. The conventional Smith predictor cannot capture the evolution of the network states, which is the root cause of stochastic time delays.

The relationship between the delay margin and the controller gains can only help achieve a compromise between the LFC performance and the maximum allowable delay. However, the delay margin-based method cannot compensate for random delays because it does not involve the real-time prediction of delays in each period. Therefore, in Chapter 3, the Smith predictor estimates the real-time delay, which is then integrated into the delay margin-based method to enhance the frequency stabilization performance. The Smith predictor can also enhance the LFC performance of robust proportional-integral-derivative (PID) controllers, whose gains are tuned via robust evolutionary algorithms [35]. Improvements to control strategies currently implemented in real systems, e.g., delay margin-based proportional-integral (PI) controllers [17, 24] and PID controllers [35], are presented to demonstrate the effectiveness of the proposed methodology for the control of WAPSs.

The power sector is experiencing a structural trend toward decentralization, stemming from the integration of large shares of renewable energy resources (RERs) [70]. This is fostered by distributed energy resources (DERs), which require the integration of power generation means located at or near the end-user side [71, 72]. However, the stochastic nature of RERs and the load demand induces system frequency fluctuations [73, 74]. An effective control strategy is needed to maintain the system frequency at its nominal value by balancing the power generation and demand in real time. To this end, automatic generation control (AGC) schemes have been developed for damping frequency oscillations in distributed generation systems (DGSs) [74–77]. AGC is performed by computing control signals based on the system frequency and delivering balancing inputs to various energy storage systems (ESSs) to absorb (release) the surplus (deficit) power from (to) the grid [77–79]. However, the ubiquity of DERs across wide areas and the complex structure of DGSs hinder the development of dedicated communication infrastructures for DGSs with massive DERs [80–83].

Recently, AGC has been integrated with open communication networks because of its low cost, high speed, simple structure, and flexible access. Data exchanges among PMUs, generators, and the control center are provided by the open communication network in the form of time-stamped data packets [42, 76, 82, 83]. Stable AGC depends heavily on the performance of the open communication network [42, 76–78, 84–88]. Cognitive radio networks, cellular networks, local area networks (LANs), wide area networks (WANs), and wireless local area networks (WLANs) are employed as open communication infrastructures in these NCSs [79, 80, 83].

However, open communication networks are exposed to various types of degradation processes, such as network-induced time delays [77, 78, 86, 87], packet dropouts [88, 89], failures of the communication infrastructure [90], uncertain communication links [91], and cyberattacks [92]. As a result, the measurement signals (control signals) received by the control center (ESS or generators) degrade, effective AGC cannot be carried out, and the system frequency response worsens [78–82]. Studying the performance of open communication networks is critical for understanding the occurrence of time delays and packet dropouts. To this end, medium access and packet transmission must be analyzed. The MAC layer is the lower layer of the data link layer of the Open System Interconnection model, and it is responsible for moving data packets among network interface cards across communication channels. Several MAC protocols, for example, carrier-sense multiple access with collision detection (CSMA/CD, Ethernet), CSMA with arbitration on message priority (controller area network), and IEEE 802.11b/g (WLAN), prevent the collision of packets sent from different nodes across the same channel [83, 93–95].

Time delays are variable, challenging to predict, deteriorate AGC performance, and reduce the stability region [78, 79]. Packet dropouts refer to lost messages, which occupy network bandwidth but cannot reach the destination. They affect the operations of DERs and the reduction of frequency fluctuations, particularly in uncertain network environments. Optimal feedback AGC regulators for DERs have been investigated in numerous works for perfect communication networks, and the impact of transmission delays and packet dropouts on the controller cannot be captured [96]. Robust PID controllers against constant or uniformly distributed time delays [77–80] are designed to cope with perturbations of the control parameters. However, constant or uniformly distributed time delays cannot be generally assumed in realistic communication networks.

In addition, recent studies focusing on primary and secondary control levels have been extended to the power management level by considering fuzzy controllers [97, 98], decentralized power management and sliding mode control strategies [99], static synchronous compensators [100], and two-degrees-of-freedom feedback-feedforward robust controllers [101, 102]. The reactive power reference can be determined and controlled by a novel application of radial basis function neural networks [103–105] to improve the power sharing and stability of microgrids with multi-DERs. To provide high reliability and robustness against network failure or time delays, droop-based control schemes are designed to specify the frequency of each DER unit by using complementary loops and fuzzy logic controllers [107], robust H_∞ controllers [86], and PI controllers [107, 108]. On the other hand, novel approaches for mitigating the impact of random time delays quantify robust delay margins [109]. The delay margin-based sparsity-promoting wide-area control strategy, which requires few system observations, can reduce communication requirements and yield nearly optimal performance compared with centralized control [109]. Nevertheless, packet dropout still has the potential to affect the performance of this strategy.

1.2.2 Reliability of CPSs

Distributed renewable energy sources are increasingly connected to power distribution networks as a remedy for environmental and economic concerns [110–112]. However, their power outputs are dependent on the available intermittent natural resources, such as solar irradiation, wind velocity, and biofuel production [113–115]. The rapid deployment and commercialization of storage devices and electric vehicles (EVs) has become an attractive technological solution to facilitate the use of renewable energy sources, manage demand loads, and decarbonize the residential sector [115–117]. The above technological issues call for managing real-time energy imbalance in DGSs to meet electricity demand over a long-term horizon. In order to address the challenges of distributed control of energy sources, communication networks are being installed for accurate control of the different power sources and the timely operational scheduling of distributed generator (DG) units, with the objective of providing reliable and sustainable energy in a timely fashion [118–123]. However, most existing research works do not formally investigate the capability of communication networks in providing real-time power management and promoting the optimal power dispatch [124–127]. The effective integration of communication networks into DG systems is a key step in the realization of future smart grids [90, 128].

Most integrated system-of-systems models have been developed based on dedicated and closed communication networks, where the infrastructure is exclusively built for smart grid applications [90, 128, 129]. As the network is dedicated between the DG and the control center, the data exchange is assumed to be perfect and free of defects (e.g., induced time delays and packet dropouts [130–134]). However, experience has shown that dedicated communication networks are ill-suited to future DG systems, which require a different, more complex but much cheaper network, as its dimension would be much larger [135–137]. Because of the low installation cost, high transmission speed, and flexible access, the open communication network has the highest potential for integration with future DGSs [82, 84, 138]. As end-users have to share the limited bandwidth in the open communication networks, which could lead to local congestion, they can be unreliable and suffer from network-induced delays and packet dropouts [139–141].

Existing research works [42, 123, 139–146] do not model explicitly and adequately the behaviors of transmission delays and packet dropouts. Most of the aforementioned models are limited to constant or less stochastic transmission delays, which are not true in reality [147, 148]. The delays are described by discrete-time models or are neglected by assuming that they are much smaller than the communication interval [17, 123, 149, 150]. Packet dropouts are usually modeled by a two-state Markov chain and the associated quantitative loss rates; the detailed state evolution is masked and only input/output information is made available. The state transition matrix is known by assuming that the evolutions of packet dropouts can be fully observed [123, 140]. Additionally, the models of uncertain renewable power sources do not consider time-correlated properties [42, 123, 139, 142, 143, 151]. Consequently, the control schemes derived based on such assumptions can be very conservative and may not be readily applicable to real systems.

To bridge this gap, a generic and transparent mathematical model is necessary for the analysis of the impacts of integration of unreliable open communication networks (e.g., home area networks, neighborhood area networks, and WANs). The major challenges lie in the modeling and simulation of the interactions between degraded communication networks and DG systems, and the optimization of the real-time energy management problem on such system platforms. Note that the specific requirements (i.e., high-frequency data and data prediction) introduced by communication network integration need to be taken into account in this modeling, simulation, and optimization framework.

1.3 Opportunities for CPS Applications

1.3.1 Managing Reliability and Feasibility of CPSs

CPSs perform critical tasks in many industrial applications, for example, manufacturing systems [152], transportation systems [153, 154], and power systems [18, 155]. The components of control systems, that is, actuators and sensors, are subject to degradation when operating under severe working conditions [155–159].

Sudden load variations in electric power systems are often balanced by promptly changing the output of natural gas power plants following the LFC strategy [160]. However, the degradation of gas turbine compressors, that is, the deviation of compressor flow capacity and isentropic efficiency [161], and the degradation of PMUs, that is, measurement drifts and errors [162, 163], reduce the LFC performance by decreasing the available balancing power, and by producing inaccurate frequency readings. Deteriorated LFC performance may result in power system failures because the system frequency exceeds its maximum allowable drop or fails to attain the steady-state frequency tolerance band in the required time in compliance with ISO 8528-5 [164]. As a result, power system failures are determined by the LFC performance, stemming from the partial information on the power system conditions, that is, the health indexes of gas turbines (flow capacity and isentropic efficiency), and measurement drift. Therefore, it is necessary to study these degradation processes to predict the real-time LFC performance loss and ensure adequate LFC through proper maintenance activities.

Lifetime prognostic studies of gas turbines [165] have shown that the failure time of these systems varies between 24 000 h and 35 000 h. Such variability stems from different working conditions [166] and different starting points of the degradation paths [167]. Therefore, the degradation model should reflect the unit-to-unit variability in the degradation process of gas turbines [168]. Moreover, many condition-based maintenance (CBM) models determine maintenance activities based on the estimated remaining useful lifetime (RUL) of gas turbines using health indexes (reduction in flow capacity and isentropic efficiency [161]) derived from on-line measurements, for example, rotor speed, inlet temperature, and pressure [169, 170]. Maintenance activities are scheduled based on the two health indexes. However, in control systems, failures are determined by the system performance [171] and not by the health indexes of individual components, that is, they are system-level failures rather than component-level failures [172–174]. Control systems may still operate normally even if the component health indexes exceed failure thresholds [157, 175].

In addition, the feedback control mechanisms hide the explicit mapping from the individual component degradation state to the control performance loss [175]. It is easy to estimate the RUL of an individual component, but difficult for degraded control systems, because their failure time cannot be described by any particular distribution [171, 176]. Therefore, when applied to degraded control systems, general CBM models should connect the component degradation and the control performance loss, so that maintenance activities depend on the reduction of control performance [176].

The Wiener process is often employed in degradation models because of its favorable mathematical properties; in particular, it can capture non-monotonic degradation signals frequently encountered in practical applications, because consecutive independent increments are normally distributed [172]. Therefore, this stochastic process has been widely applied to characterize the path of degradation in realistic scenarios, where fluctuations are observed in the degradation process, for example, brake-pad wear for automobiles [177], bearing degradation [174, 178], gyroscopes in inertial navigation systems [172, 179, 180], contact image sensors in copy machines [181], the resistance of carbon-film resistors [159], and the pitting corrosion process [167]. To overcome the aforementioned limitations, the Wiener degradation model with unit-to-unit variability is introduced to describe gas turbines exhibiting different lifetimes. The Wiener model considers the random starting time of the degradation process, which follows a non-homogeneous Poisson process [167, 182]. Furthermore, the drift parameter, which denotes the aging rate, is also variable and follows a normal distribution, where the mean denotes the average aging rate and the variance represents the variability in the aging rate [170, 183–186]. These parameters can be estimated from the lifetime dataset via an expectation-maximization (EM) algorithm [172, 180, 183–186].

Thus, the data-driven degradation model with unit-to-unit variability is integrated into the control system model described by control block diagram, resulting in a real-time simulation model. In such control models, the interplay among the reduction in the control signal due to component degradation, the transfer functions of the subsystems, and the feedback control loop, provides the mapping between the component degradation states and the system performance loss. This interaction is modeled via control-block diagrams, which implement the feedback control mechanism and quantify the control signal by comparing the control performance to the setpoint. Therefore, such an integrated model does not require explicit mapping from the component degradation states to the system performance loss and is well-suited to represent a degraded control system. As such, this simulation model realistically predicts the performance of the control system at different operating times and degradation stages.

1.3.2 Ensuring Cybersecurity of CPSs

Attacks on complex systems, for example, CPSs, are fundamentally different from traditional internal failures (e.g., degradation and design) and external failures (e.g., natural disasters) [187–190]. Many attack models for complex systems embrace a partial perspective, which only focuses on component vulnerability, and neglects the dependence of system performance on it [191–193]. As a result, the insights provided by these models are not adequate for providing general recommendations in realistic applications. To address this limitation, recent studies investigate the influence of component vulnerability (attacks at the component level) on system performance [194–197].

Pioneering works [192, 198–202] develop optimal defense strategies to minimize the attachment vulnerability of parallel systems, assuming that attackers maximize either the damage probability or the expected damage over a time horizon. They also consider general features, that is, imperfect false target techniques and genuine targets [201, 203]. These defense strategies reach a trade-off between increasing the protection of existing components and providing redundancy by allocating additional components [192, 203–205].

System performance is an essential feature in CPSs that can still operate if some components are unavailable and, therefore, are characterized by multiple performance levels [206–211]. System performance degrades with increasing component destruction or unavailability; if the system performance level decreases, the required demand may be partially unsatisfied. Two risk measures can be used for multi-state complex systems [203–205]: 1) the probability that the demand is not satisfied is considered for complex systems that fail if performance cannot meet demand, for example, automatic train protection and block systems [212, 213], and power system dynamic security systems [190]; 2) the expected damage proportional to the unsupplied demand is considered for complex systems that can operate even if the demand is partially supplied, for example, mobile ad hoc networks [191], NCSs [214, 215], supervisory control and data acquisition (SCADA) systems [216, 217], water distribution networks [218], and electric power grids [219–221].

Several works consider both the vulnerability and performance of complex systems subject to attacks [201–207, 222, 223]. These works generally describe a case as a dynamic contest between an attacker and a defender to develop a component vulnerability model and a multi-state system performance model. The number of destroyed components quantifies the demand loss and expected damage costs [200, 205]. To make the above contest more realistic, attack time uncertainties and the attacker's preference on the attack time should be considered.

In the literature, two different approaches exist for determining the attack time, that is, the strategic selection and the selection based on probability distributions. In the former, the attacker strategically selects whether to attack at some point in time or at a later point in time, based on the outcome of the game, given that the attack occurs at a specific time [224]. Thus, complex attack and defense strategies can be derived from a two-stage min-max multi-period game. Extensive attack or defense in one period limits the attack or defense that can be exerted in the next period, and vice versa. Thus, players strategically choose whether to exert effort now or in the future [224–226]. The defender may determine optimal resource allocation strategies for redundancy [192] and protection, that is, individual or overarching protection [205, 227–229]. On the other hand, the attacker may distribute the constrained resources optimally across sequential attacks [230–233].

In the second approach, the attacker prefers to conduct the attack at the time of the critical event [206]. Indeed, attacks in Nice, Berlin, Manchester, and London occurred several days before and after Bastille Day, Christmas, a concert, and the Champions League 2017 Final, respectively. In these cases, the defenders have increased the protection level in the immediate aftermath; therefore, it is not worthwhile and cost-effective for the attacker to deploy another attack in a short period. Because attacks occurred at critical times, they can greatly influence public opinion. As a result, the attacker aims to maximize the system loss by strategically selecting a set of elements to attack based on the two-stage min-max game [234]. Because we can predict the distribution of the time at which the critical event occurs, the attack time can be inferred from a data-driven probability distribution [192]. The two approaches aim to maximize the outcomes of the game given that the attack occurs at a specific time under a similar system structure and variable resources.

The truncated normal distribution is used to describe the uncertainty of the most probable attack time, that is, the time of the critical events, and the accuracy of the defender's estimate of it [206, 235, 236]. The truncated normal distribution has been adopted to represent uncertainties in many realistic applications, for example, traffic peaks of online video websites [193], the peak season of power supplies [237–239], the peak demand of water distribution systems [240], and the rush hour of public transportation [241]. Accounting for the influence of this uncertainty increases the relevance of the insights gained for the optimal resource allocation strategy against attacks.

CPSs are a new class of engineered complex systems that provide tight interactions between cyber and physical components. The corruption of a small subset of their components has the potential to trigger system-level failures leading to entire system performance disruptions [191, 215, 221, 242]. Previous studies on attack vulnerability and performance of complex systems can be extended to identify resource allocation strategies for cyber components and promote system performance during cyber-attacks in CPSs [243, 244]. Cyber vulnerabilities are exploited by attackers to launch insidious attacks on the integrity, confidentiality, and availability of cyber data by injecting false data into measurement devices, eavesdropping estimation of system states, and deploying denial of service (DoS) attacks on communication networks [216, 217, 220, 245]. More sophisticated attack models specifically target weaknesses to cause maximal damage [191]. In this respect, it is key to capture the uncertainties intrinsic to the behavior of the attacker and the defender.

With respect to applications in smart grids, upgrading traditional grids to smart grids has brought many benefits to the overall management of power and energy systems, including higher reliability, better efficiency, improved integration of RERs, more flexible choice for stakeholders, and lower operation costs [246–248]. However, the core technologies, for example, communication techniques and SCADA systems [249–252], which deliver the advantages of smart grids, also open the grids to vulnerabilities that already exist in the information and communications technology (ICT) world. These vulnerabilities pose threats to smart grids, such as DoS attacks, false data injection, replay attacks, privacy data theft, and sabotage of critical infrastructure [253–255]. In addition, the failures in a smart grid caused by cyberattacks can easily cascade to other highly dependent critical infrastructure sectors, such as transportation systems, wastewater systems, health care systems, and banking systems, resulting in extensive physical damage and social and economic disruption [249, 256].

While government, the private sector, and academia are recognizing the cyber vulnerability of smart grids, the likelihood and impact of a cyberattack are difficult to quantify. Furthermore, for a smart grid, there may be mandatory standards and operational requirements from grid stakeholders. Current risk management strategies are generally qualitative or heuristic [257]. In these strategies, some assumptions, for example, constant reward with respect to successful anti-cyberattack [258, 259], may be unrealistic for most smart grids.

Chapter 7 presents a probabilistic risk analysis framework to enhance smart grid cyber security. In particular, the dynamic and stochastic characteristics of smart grids, such as uncertain demands, are taken into account to investigate the effect of defending strategies on the real operation cost. The optimal power flow (OPF) model [260] is applied to an 11-node radial smart grid originating from the Elia grid in Belgium. Compared with the existing studies that focus on the inherent risk [254, 260], such as the natural degradation and uncertain RERs for better maintenance actions and power dispatch, Chapter 7 addresses the impact of the external threat (cyberattacks) on the operation cost for effective deployment of cyber defense teams. In previous works, the cost of each attack on a node was assumed to be a constant [259]. Nevertheless, by investigating some practical scenarios, it has been found that the costs are more likely to be determined by some adversarial factors. Therefore, an adversarial cost sequence associated with each node is assumed, and a widely used variation constraint is introduced for each cost sequence. To cope with the objective of sequential decision strategies, the problem is formulated using the reinforcement learning framework [261–263]. In particular, the Bayesian prior method [259] is employed for the model parameters, and the problem is formulated as a Bayesian adversarial multi-node bandit model. In addition, a Bayesian minimax type regret function is constructed, which is subject to the learning context.

2

Fundamentals of CPSs

In this chapter, fundamental Cyber-Physical System (CPS) models, evaluation processes, verification procedures and optimization methods are introduced.

As the CPS lacks an explicit formulation due to the feedback control mechanisms [171, 176], the closed-form solution of the optimal stability, reliability and resilience strategies is unavailable. Therefore, meta-heuristics algorithms are employed to assist the search of the optimal strategy. For maintenance of CPSs, we use the hybrid Genetic-Simulated-Annealing algorithm (HGSAA), which has been applied in many maintenance optimization works [264–266]. It combines the Genetic Algorithms (GA) and the Simulated Annealing (SA) to improve the quality of solutions and reduce computation efforts [266]. For the optimal control strategy of CPSs (single objective), A particle swarm optimization (PSO) method based on Monte Carlo Simulation (MCS) is introduced in this book to solve the Proportional–Integral–Derivative (PID) controller optimization problem [15, 267]. The method can achieve higher search efficiency values since it combines local search with global search. The technique has been widely employed to find optimal solutions for realistic applications, especially with regard to the optimization of control systems [268–270]. A solution including the PID control strategy is encoded as a finite-length string called "a particle" in PSO. The fitness value of each particle is determined by a fitness function. However, the optimization of PID usually involves simultaneous optimizing two objectives, i.e., control performance and control effort, which are generally non-commensurable and conflicting with each other [271, 272]. Better control performance means more control efforts and vice versa. Therefore, the multi-objective PSO algorithm is applied to achieve a best trade-off between the two objectives [273–275].

2.1 Models for Exploring CPSs

2.1.1 Control-Block-Diagram for CPSs

Consider the CPS with degraded components shown in Figure 2.1, which is a typical case of CPSs consisting of a forward channel and a feedback channel.

Part of the contents are from [175, 180, 270, 293, 294] and permission has been obtained to use the contents.

Cyber-Physical Distributed Systems: Modeling, Reliability Analysis and Applications, First Edition.
Huadong Mo, Giovanni Sansavini and Min Xie.
© 2021 John Wiley & Sons Ltd. Published 2021 by John Wiley & Sons Ltd.

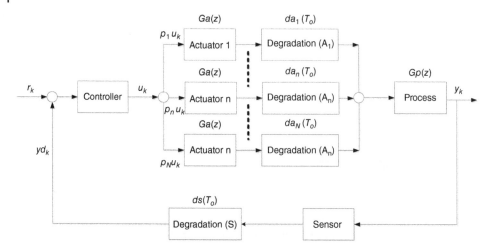

Figure 2.1 The control block diagram of CPSs with degraded components.

2.1.1.1 Control Signal in CPSs

At the total run time T_o, for an operational task with total K sampling periods, the time-varying model of the forward channel in the kth period ($k \in \{1, ..., K\}$) is first derived.

As the sensor measures the system output during every sampling period T_P, the control signal u_k remains the same during the interval $[(k-1)T_P, T_P]$, which is the inherent discrete-time property. Thus, the control signal u_k can be represented by the following sum of input steps:

$$u_k = u_1 + (u_2 - u_1)I(t - T_P) + \cdots + (u_k - u_{k-1})I(t - (k-1)T_P), \ t \geq 0 \tag{2.1}$$

where $I(t) = \begin{cases} 1, & \text{if} \quad t \geq 0 \\ 0, & \text{else} \end{cases}$.

Arriving at the relationship between control signal u_k and output y_k is not a trivial task. In the time domain, it cannot be represented by the simple multiplication of time-varying model of the actuators, their respective degradations, and the process, but their convolution. Inspired by a similar problem solved by control theory, by applying the Laplace transform to respective time-varying models of the actuators, degradations, and the process, their respective complex domain models can be obtained. Once this is done, the overall relationship can be described through simple multiplication of the complex domain models.

Firstly, applying the Laplace transform to u_k, its complex domain representation is:

$$U_k(s) = u_1 + (u_2 - u_1)\frac{e^{-T_P \cdot s}}{s} + \cdots + (u_k - u_{k-1})\frac{e^{-(k-1)T_P \cdot s}}{s} \tag{2.2}$$

where $U_k(s) = \mathcal{L}(u_k)$ with $u_k = [u_k^1, u_k^2, \ldots, u_k^N]$.

2.1.1.2 Degraded Actuator and Sensor

Since the system has N identical actuators, for any $n \in \{1, 2, \ldots, N\}$, the degradation of the nth actuator can be modeled by a Wiener-process subject to unit-to-unit variability [276, 277]:

$$da_n(T_o) = da_n(0) + \theta T_o + \sigma B(T_o) \tag{2.3}$$

where $da_n(T_o)$ denotes the loss in the effectiveness of the nth actuator at total run time T_o. $da_n(0)$ is a known initial degradation state. Without loss of generality, it is assumed that $da_n(0) = 0$ in this book. For any $T_o > 0$, $\sigma B(T_o) \sim N(0, \sigma^2 T_o)$ describes the stochastic dynamics of the degradation process.

It is found that (2.3) is able to describe the time-varying variability arising from the dynamic features of $B(T_o)$. Nevertheless, since each actuator operates under different working conditions, different actuators follow different degradation paths at different rates. Compared with previous works where the degradation processes are identical, it is more realistic to consider the unit-to-unit variability in the degradation process. Thus, θ is regarded as a random parameter denoting the unit-to-unit variability with $\theta \sim N(\mu_\theta, \sigma_\theta^2)$, and σ is regarded as a constant capturing the nature of the degradation process common to all actuators in the same class.

Assumption 2.1: For any $kT_P \ll T_o$, $da_n(T_o + kT_P)$ equals to $da_n(T_o)$. Thus, $da_n(T_o + kT_P)$ is a constant value during an operational task starting at total run time T_o. The values of θ and $B(T_o)$ are independent of each other.

Since the operational time of one control process is much smaller than the total run time of entire CPSs in real applications, it is general to consider that the change in the degradation of one component is negligible in such a short time [13, 278]. Therefore, above assumption is reasonable and not restrictive.

Since the degradation of each actuator is constant during an operational time interval $[T_o, \ldots, T_o + KT_P]$, the Laplace transform cannot be applied to a constant value. Even so, the relationship between control signal and system output can still be represented by simply multiplying the time-varying models of the actuators and the process and the degradation constant as:

$$Y_k(s) = U_k(s)Ga(s)[I - da(T_o)]Gp(s) \tag{2.4}$$

where $Y_k(s) = \mathcal{L}(y_k)$ and $da(T_o) = [da_1(T_o), \ldots, da_N(T_o)]^{\text{tr}}$, where tr stands for the transpose of matrix.

In industrial applications, the perfect sensor measurement of system output state is often impossible. It is the common case for the controller that the received sensor measurements are not only subject to measurement error but also to sensor gain degradation due to the noise and severe working conditions [279, 280]. Therefore, the received measurement in the kth period is a superposition of the measurement error w_k and sensor gain degradation $ds(T_o + kT_P)$ with

$$ds(T_o) = ds(0) + \lambda T_o + \sigma B(T_o) \tag{2.5}$$

where the $ds(T_o)$ accounts for the stochastic sensor gain degradation at total time T_o, $ds(0)$ is the initial state with $ds(0) = 0$, and λ is the drift coefficient with $\lambda < 0$. Compared with the aforementioned models in which $ds(T_o)$ is restricted to be a constant or follow a uniform or Bernoulli distribution, (2.5) is more suitable to describe the time-varying properties of the sensor gain degradation.

Assumption 2.2: For any k and $kT_P \ll T_o$, $ds(T_o + kT_P)$ equals to $ds(T_o)$ during an operational task starting at time T_o. $ds(T_o)$, and w_k and $da(T_o)$ are independent of each other.

Thus, the measurements received by the controller can be described as

$$y_k = [1 - ds(T_o)]y_k + w_k \tag{2.6}$$

2.1.1.3 Time-Varying Model of CPSs

Therefore, substituting (2.2) into (2.4), and using the linearity property of Laplace transform and applying the inverse Laplace transform to (2.4), the output y_k can be obtained as

$$y_k = \mathcal{L}^{-1}\left\{\left[u_1 + (u_2 - u_1)\frac{e^{-T_P \cdot s}}{s} + \cdots + (u_k - u_{k-1})\frac{e^{-(k-1)T_P \cdot s}}{s}\right] Ga(s)[I - da(T_o)]Gp(s)\right\}$$

$$= [u_1 I(t)F_1(t) + (u_2 - u_1)I(t - T_s)F_2(t - T_P) \tag{2.7}$$

$$+ \cdots + (u_k - u_{k-1})I(t - (k-1)T_P)F_2(t - (k-1)T_P)] \cdot [I - da(T_o)]$$

where $F_1(t) = \mathcal{L}^{-1}(Ga(s)Gp(s))$ and $F_2(t) = \mathcal{L}^{-1}(Ga(s)Gp(s)/s)$.

After completing the forward and inverse Laplace transform, the time-varying model of the forward channel is built. Consider now the time-varying model of the feedback channel.

Right after the controller receives the measurement of the system output y_k in (2.6), it will compute the control signal for each actuator at the $(k + 1)^{\text{th}}$ period, separately. For simplification, it is assumed that all actuators share the control signal evenly.

Thus, the control signal for actuator n based on the following PID strategy is determined as:

$$u_{k+1}^n = K_p\left[e_{k+1} + \frac{T}{T_i}\sum_{j=1}^{k+1} e_j + \frac{T_d}{T}(e_{k+1} - e_k)\right] / N \tag{2.8}$$

where $e_{k+1} = r_{k+1} - y_k$.

From (2.8), it can be concluded that the control signal computed is determined actually by historical errors. Properly designing the parameters of the control strategy can definitely help the system eliminate the influence caused by degraded components and make the system output converge to the expected target with required qualities.

Therefore, the stochastic model of CPSs is closed through the forward channel model expressed by (2.1) and (2.7), and the feedback channel model expressed by (2.6) and (2.8). The performance analysis and reliability improvement can be conducted based on the explicit system structure by optimally choosing the parameters of the system controller.

2.1.2 Implementation in TrueTime Simulator

2.1.2.1 Introduction of TrueTime Simulator

The architecture for the Automatic Generation Control (AGC) of Distributed Energy Resources (DERs) via Ethernet and hybrid network, is illustrated by the Figure 2.2. To implement the microgrid with two different communication network architectures in the Matlab/Simulink environment, the TrueTime simulator is deployed for simulating different types of communication protocols [281].

TrueTime is a Matlab/Simulink-based simulator for real-time CPSs, which facilitates co-simulation of controller task execution in real-time kernels, network transmissions, and continuous plant dynamics. It can simulate most realistic communication networks, e.g., Ethernet, CAN, TDMA, FDMA, Round Robin, Switched Ethernet, FlexRay, PROFINET, 802.11b WLAN and 802.13.4 ZigBee. Multiple activity levels of the interfering traffic are simulated via the Interference node, which sends random interfering packets over the network. Packet dropout is described by Bernoulli-distributed variables [282]. More details of TrueTime simulator and its realization in CPSs are given in this Section.

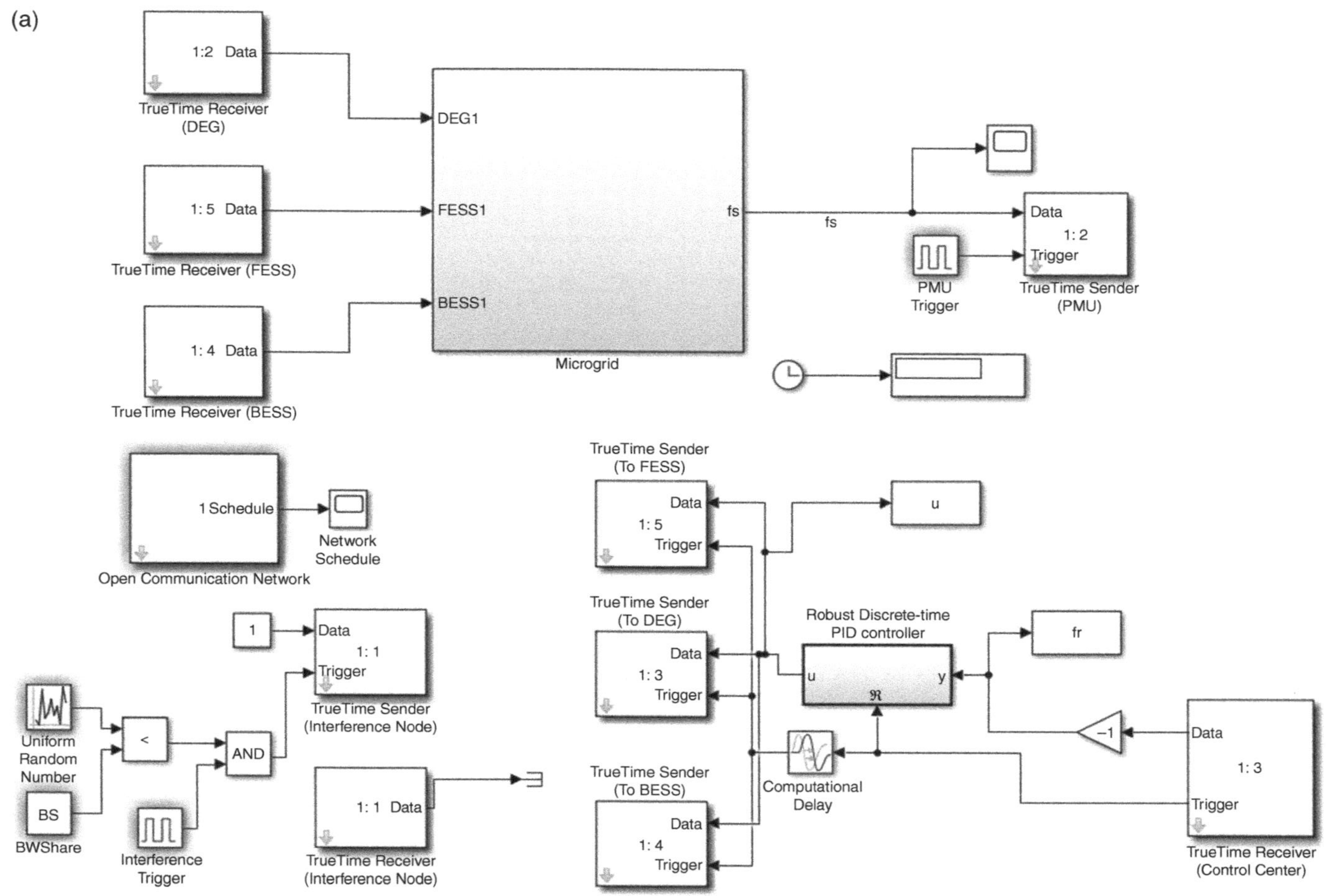

Figure 2.2 Simulink realization of the microgrid. (a) Ethernet, (b) Hybrid network.

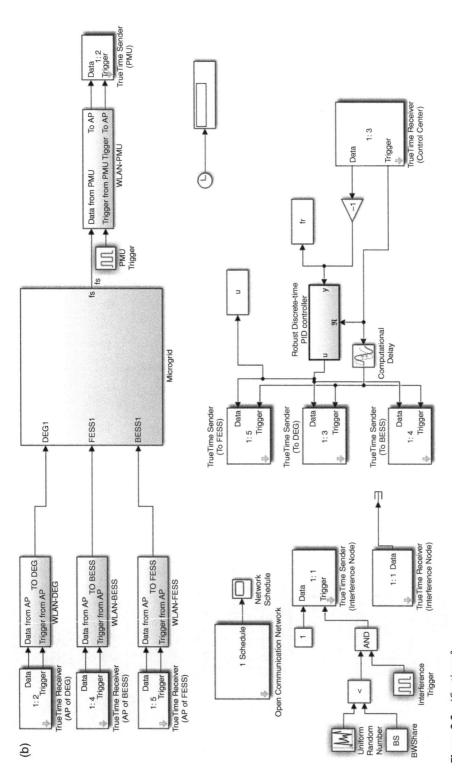

Figure 2.2 (*Continued*)

Above procedure is detailed in the "TrueTime 2.0-Reference Manual", which provides representative examples of CPSs [283]. Figure 2.2 (a) and Figure 2.2 (b) present the Matlab/Simulink implementations of the microgrid with Ethernet and of the microgrid with Hybrid network, respectively. The network blocks from the TrueTime simulator are directly linked with physical components, i.e., the PID controller, Phasor Measurement Units (PMU), DERs, to enable data exchanges.

The data exchange among the control center, DERs, interference node and PMU are wired in the Ethernet architecture. In the hybrid architecture, the data exchange between the routers and the Remote Terminal Unit (RTU), such as Battery Energy Storage Systems (BESS), Flywheel Energy Storage Systems (FESS), Diesel Engine Generator (DEG) and PMU, is wireless and provided by the 802.11b/g. The data exchange among routers is wired and provide by the Ethernet. Low product prices make 802.11b/g more convenient and cost-effective compared to the Ethernet. As such, 802.11b/g has become an efficient approach to provide flexible data communication between routers and RTU, and is employed for monitoring and controlling DERs, offshore wind farms and smart home energy management systems [80].

The interference node simulates the interference user in the open network, who sends disturbing traffic over the channels and cause congestion. Generally, the length of data traffic is constant (i.e., 80 bytes in this chapter, same as the length of PMU measurement [283, 284]), and the interference node sends it to the network at every time period T_i if

$$UN_i < BWShare \tag{2.9}$$

where UN_i is a uniformly distributed random number sampled at time period T_i in the interval $[0, 1]$, and BWShare is the expected ratio of the network bandwidth used by the interference node [283].

Remark 2.1: The time interval T_i determines the activity level of the interference user. For example, $T_i = 0.01$ s means that the interference user tries to send out disturbing traffic every 0.01 s.

Remark 2.2: BWShare indicates the expected percentage of bandwidth used by the interference user. If the node sends out a message, it occupies the entire channel bandwidth, and other nodes cannot send messages until this channel is free, to avoid packet collisions.

For example, BWShare=0.1 means that 10% of bandwidth is used by the interference user, and from a statistical viewpoint, the interference user sends out disturbing traffic with probability 0.1.

2.1.2.2 Architectures of CPSs in TrueTime

The architectures of the real-time AGC of DERs via the Ethernet and hybrid network all consists of:

- PMU node (time-driven): the PMU takes measurements of the system frequency at every sampling interval $T_s = 0.01$ s and sends $\Delta f(kT_s)$ to the control center over the network. The size of data packets is 80 bytes and the phase delay caused by the filter of the PMU is 0.006 s [284, 285].
- Control center node (event-driven): when a data packet from the PMU reaches the control center, the controller takes 0.002 s to compute the control signal $u(t_n)$ and sends it to the DERs. The length of control signal is 500 bytes.

- DERs nodes (event-driven): the DEG, BESS and FESS adjust their operations based on the control signal received.
- Interference node (time-driven): it sends disturbing traffic over the network with period T_i and causes congestion, to generate different scenarios of time delay and packet dropout.

The above two architectures for the open communication network are simulated using the TrueTime simulator [283], which is a Matlab toolbox that generates scenarios of realistic network with different MAC protocols [43, 83, 283].

Example 2.1: Figure 2.3 presents the dynamics of sending and receiving time for data packets at the interference node, PMU, control center and DERs under the Ethernet architecture, for BWShare = 0.3, $T_i = 0.005s$ and $L_d = 5\%$. The continuous lines show steps when a message is received by the corresponding node. The x coordinate of the steps identifies the receiving time of data packet, and the y coordinate identifies the time at which that corresponding packet was sent. Dotted vertical lines indicate that one of the nodes, i.e., interference node, PMU and control center, respectively in Figure 2.3, sends out a data packet. The sending time marked by the dotted lines is consistent with the y value of step.

In Figure 2.3 (a), the step (55, 46) in the control center shows that the PMU sends out a system frequency measurement at 46 ms, which is consistent with the dotted line showing a sending pulse from the PMU at 46 ms, and the control center receives it at 55 ms. Furthermore, Figure 2.3 allows identifying the dropped packets. In Figure 2.3 (d), the control center sends out a packet to FESS at 29 ms (dotted lines marked with cross), but the FESS cannot receive it (no corresponding step with y = 29 ms), indicating that a packet dropout occurs in this transmission. Above two scenarios are highlighted in the Figure 2.3 to provide better descriptions.

2.2 Evaluation and Verification of CPSs

2.2.1 CPS Performance Evaluation

2.2.1.1 CPS Performance Index

Having built the hybrid model of the closed-loop feedback CPS with degraded components, an event-based MCS method is introduced to quantitatively assess the reliability of such system. The quantitative analysis for system performance considers the domain performance requirements [286].

The qualities of a control process of CPSs consist of the rising/declining time, the percentage overshoot and the settling time, which are all relevant in assessing real-time operational performances. Descriptions of the qualities are presented in Table 2.1. Rising/declining time is useful while ensuring that the control process responds rapidly to the command signal; percentage overshoot ensures that the control process exhibits sufficiently small oscillations during response; and the settling time ensures that the control process quickly reaches the expected output value and keeps stable.

If any of the quality parameters exceeds the corresponding maximal operational requirements within the total run time T_o, it can be concluded that the CPS is unable to satisfy the

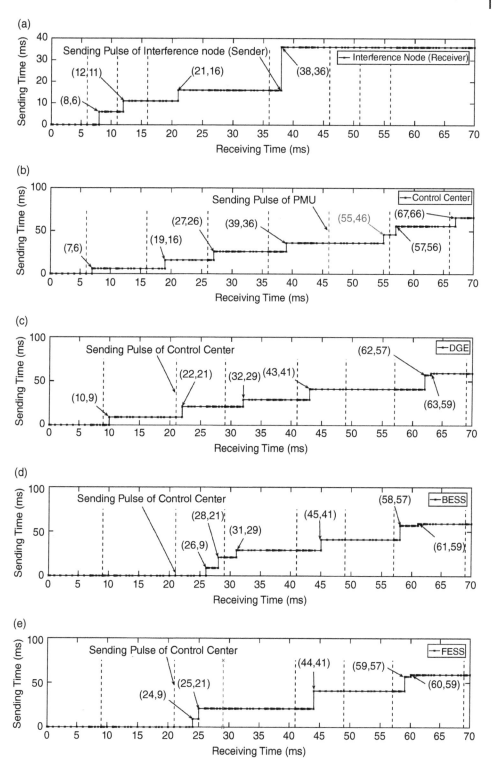

Figure 2.3 Sending time and receiving time for the (a) interference node, (b) control center, (c) DGE, (d) BESS and (e) FESS in the Ethernet architecture, where BWShare = 0.3, $T_i = 0.005s$ and $L_d = 5\%$.

Table 2.1 Domain Requirements and Descriptions.

Domain requirements		Descriptions
Operational	Rising/declining Time RT/DT	The time taken by the output of the control process to rise/decline from a low/high specified value to a high/low specified value.
	Percentage Overshoot PO	The absolute value between the expected output of the control process and the highest/lowest output divided by the expected output.
	Settling Time ST	The time elapsed from the application of the control signal to the time at which the output of the control process has entered and remained within a specified error band.
Nonfunctional	Reliability	The ability of the control system to maintain the expected performance in the presence of the degraded components.

stipulated requirements. Hence, it has failed at total run time T_o, because of component degradation.

When considering the quality-Rising time, it can be represented as

$$P(x = 0) = P(RT > RT_{max} \cup PO > PO_{max} \cup ST > ST_{max}) \tag{2.10}$$

When considering the quality-Declining time, it can be represented as

$$P(x = 0) = P(DT > DT_{max} \cup PO > PO_{max} \cup ST > ST_{max}) \tag{2.11}$$

where RT, DT, PO and ST are the qualities of the control process at total run time T_o. $x = 0$ means that the CPS has failed at total run time T_o. RT_{max}, DT_{max}, PO_{max} and ST_{max} are the maximal operational requirements.

Besides the operational requirements, the nonfunctional requirement-reliability is especially important. Reliability directly determines the ability of the CPS to maintain the expected qualities despite the presence of degraded components.

2.2.1.2 Reliability Evaluation of CPSs

It is noticed that the degradation process associated with each component is probabilistic and it is difficult to obtain the explicit reliability function of such a system. The event based MCS method is introduced in this book to estimate the ability of the system with a PID control strategy in the design phase [13, 15]. The reliability of the CPS is then estimated as a tabulated function of the total run time subject to operational requirements. The detailed procedures for applying MCS method to hybrid model are described below.

INPUT: the PID control strategy K_p, T_i and T_d; the mathematical model for the CPS; the degradation paths of components; the total run time T_o; the operational requirements.

OUTPUT: the estimate $R^*(T_o)$, of system reliability at total run time T_o.

STEP 1: Set current number of MCS replications h equal to 1.

STEP 2: Define precision interval L, the percentage of simulations belonging to this precision interval p, and the number of simulations N_T in each MCS run; set N_F, N_E and current number of simulations i in this MCS run to 0.

STEP 3: Conduct a simulation using the hybrid model described in section 2.1.1.3. If the simulation fails as judged by (2.10) or (2.11), then set $N_F = N_F + 1$, the current reliability $R_i^*(T_o) = 1 - \frac{N_F}{i}$ and note the qualities of process RT_i/DT_i, PO_i and ST_i.

STEP 4: If $|R_i^*(T_o) - R_{i-1}^*(T_o)|/R_{i-1}^*(T_o) \leq L$, $N_E = N_E + 1$.

STEP 5: If $\frac{N_E}{i} < p$ or $i < N_T$, let $i = i + 1$ and go to **STEP 3**; otherwise, $R_h^*(T_o) = 1 - \frac{N_F}{N_T}$ and go to **STEP 6**.

STEP 6: If $h < H$, let $h = h + 1$ and go to **STEP 2**; otherwise, $R^*(T_o) = \sum_{h=1}^{H} \frac{R_h^*(T_o)}{H}$ and halt.

The estimated reliability $R^*(T_o)$ has the following several important statistical properties which have been proved in literature [15, 28, 76]. These properties are rather useful in verifying whether the estimated reliability of system obtained from the event based MCS is a good estimator and defining the relationship among H, ε and α. The unbiased estimator with required precision derived from a small number of simulations is important for reliability assessment. Several properties about the estimated reliability are introduced as follows.

Property 2.1: The expectation of the estimated reliability $R^*(T_o)$ of the CPS with degraded components obtained from MCS method is an unbiased estimator of the exact reliability $R(T_o)$ at total run time T_o.

Proof: Let $R_h^*(T_o)$ be the expected reliability of the CPS with degraded components at total run time T_o, where $h = 1, 2, ..., H$. Then,

$$E[R_h^*(T_o)] = 1 \times P(x = 1) + 0 \times P(x = 0) = R(T_o) \tag{2.12}$$

Thus,

$$E\left[\sum_{h=1}^{H} \frac{R_h^*(T_o)}{H}\right] = \sum_{h=1}^{H} \frac{E[R_h^*(T_o)]}{H} = \sum_{h=1}^{H} \frac{R(T_o)}{H} = R(T_o) \tag{2.13}$$

Property 2.2: $R^*(T_o)$ is an unbiased, consistent estimator of $R(T_o)$:

$$Var[R^*(T_o)] = \sum_{h=1}^{H} \frac{R(T_o)[1 - R(T_o)]}{H} \tag{2.14}$$

Property 2.3: If the absolute error ε and the confidence level $(1 - \alpha)\%$ of the MCS method are required, the total number of MCS replication H satisfies below approximate relation

$$H \geq (Z_{\alpha/2} \cdot S_0/\varepsilon)^2 \tag{2.15}$$

Proof: Let $R^*(T_o)$ be the estimated reliability and $R(T_o)$ be the actual reliability. Consider the absolute error ε and the confidence level $(1 - \alpha)\%$, it exists

$$P(|R^*(T_o) - R(T_o)| \leq \varepsilon) \geq (1 - \alpha)$$

Using $Z_{\alpha/2}$ as the z-value of the $(1 - \alpha/2)$ percentile of the standard normal distribution, set the initial estimate of the replications required for MCS as the smallest integer H such that

$$H \geq (Z_{\alpha/2} \cdot S_0/\varepsilon)^2$$

where S_0 is the sample standard deviation of $R^*(T_o)$ obtained from a sample with H_0 MCS replications. In this book, H_0 is selected as 1000. Setting the number of MCS replications at 1500 as well as the sample size of simulations in each MCS run at 5000 is adequate for an absolute error 0.005 at the confidence level 95%. It should be noted that if H_0 is small, it would be more appropriate to use t-distribution $t_{\alpha/2, H_0}$ instead of $Z_{\alpha/2}$.

A high reliability indicates that the designed CPS should be able to satisfy the operational requirements in most cases while considering the degradation of the components. Otherwise, the system should be redesigned by optimizing the PID control strategy, in order to improve the ability of the feedback system compensating the loss effectiveness caused by degraded components, and finally provide a required reliability.

2.2.2 CPS Model Verification

The validation of data-driven degradation model of the component in the CPS is usually via comparison and this chapter takes the model of generator degradation as the example to show how to conduct the cross-verification with existing estimation model.

Example 2.2: The power system with a natural gas power plant consisting of four gas turbine units is shown in Figure 2.1. The values of the power system parameters are given in [18, 160], i.e., $b = 1, c = 0.05, X = 0.6, Y = 1, T_c = 0.3, T_f = 0.23, T_t = 0.2, K_p = 120, T_p = 20$, $B = -0.425$ and $R = 2.2$. The discrete PID controller for the LFC is optimally tuned via the two-degree-of-freedom internal model control design method and the PID approximation procedure [18].

This tuning method provides the system with disturbance rejection ability, and results in $K_P = 0.13$, $K_I = 0.42$ and $K_D = 1.32$, which ensure optimum LFC performance ($PO_O = -0.015$ pu and $ST_O = 2.5$ s), against the sudden load increase and the requirements in [264], i.e., failure thresholds $L_{PO} = -0.016$ pu and $L_{ST} = 30$ s. The steady output of the gas turbines is $SO_O = 2.5 \cdot 10^{-3}$ pu.

The degradation of the gas turbines and of the PMU reduces the LFC performance and ultimately results in the failure of the power system [161, 165, 266]. For simplicity, time T, λ and σ_B are expressed in units of 10000 h in this example, unless other units are specified. The degradation rates of the compressors flow capacity and efficiency are estimated by applying the EM algorithm to the dataset in [161]:

- the flow capacity degradation of gas turbine is described by: $\lambda \sim N(1.8568, 0.0473^2)$, $\sigma_B = 0.1, a = 1$;
- the efficiency degradation of gas turbine is described by: $\lambda \sim N(1.0252, 0.0427^2)$, $\sigma_B = 0.055, a = 1$.

The model parameters for the measurement drift and the measurement error of the PMU ($T_s = 0.1$ s) are given in [162]:

- the measurement drift degradation of sensor is described by: $\lambda \sim N(1.1575, 0.0605^2)$, $\sigma_B = 0.04$, and $a = 1$ [161];
- the measurement error is described by: $w[mHz] \sim N(0, 2^2)$.

In addition, Table 2.2 presents the parameters of the distribution of the aging rate, i.e., the drift parameter λ, obtained by the EM algorithm and the 95% confidence interval.

Figure 2.4 (a) and Figure 2.4 (b) present the comparisons among the degradation models used in [161, 288], and the degradation model of (2.3) for two gas turbines.

Table 2.2 The parameters and the 95% confidence interval of the distribution of the drift parameter λ.

Case	Parameter	95% Confidence Interval
Flow Capacity	$u_\lambda = 1.8568$	[1.8526 1.8609]
	$\sigma_\lambda = 0.0473$	[0.0445 0.0504]
Efficiency Degradation	$u_\lambda = 1.0252$	[1.0215 1.0290]
	$\sigma_\lambda = 0.0427$	[0.0402 0.0455]
Measurement drift	$u_\lambda = 1.1575$	[1.1521 1.1628]
	$\sigma_\lambda = 0.0605$	[0.0570 0.0645]

The Wiener degradation model accounting for unit-to-unit variability better represents the degradation path of the turbine health index. The mean squared errors (MSE) of the Wiener process are 0.7498 and 0.7145; the MSE of the linear and quadratic regression-based methods are, respectively, 0.8512 and 0.9665, and 0.8198 and 0.8757. Additionally, the Wiener degradation path can track the variable trend of the flow capacity drop. As such, the drift parameter λ introduces a stochastic component capturing the unit-to-unit variability.

2.3 CPS Performance Improvement

This section introduces the reliability enhancement of CPSs via using heuristic algorithms, such as GA and PSO, and the control performance improvement of CPSs via solving a multi-objective optimization problem. This section provides the basics of maintaining the performance indexes of CPSs.

2.3.1 PSO-Based Reliability Enhancement

In the present section, the PSO method is selected as an example, where a solution consisting of three parameters of the PID control strategy - K_p, T_i and T_d - is encoded as "a particle" in PSO. The fitness value of each particle is determined by the same fitness function. It should be aware of the fact that the reliability of the system may stay at 1 if the total run time T_o is in an early stage of the system's life time, which indicates that the degradation of component is not so severe.

In such a case, the PID control strategy to be adopted not only needs to make the system reliability stay at 1, but also maintain the qualities of the control process at high levels. Therefore, the fitness function should consider two cases: system reliability at 1 and not. If system reliability is 1, the fitness function is a function of qualities of the control process; otherwise, the fitness function is a function of system reliability plus a penalty for system reliability not being at 1.

Thus, the fitness value of a particle can be evaluated by the following equation:

$$f(T_o) = \varphi(R^*(T_o)) \cdot (1 - R^*(T_o) + \theta) + (1 - \varphi(R^*(T_o))) \cdot Q \qquad (2.16)$$

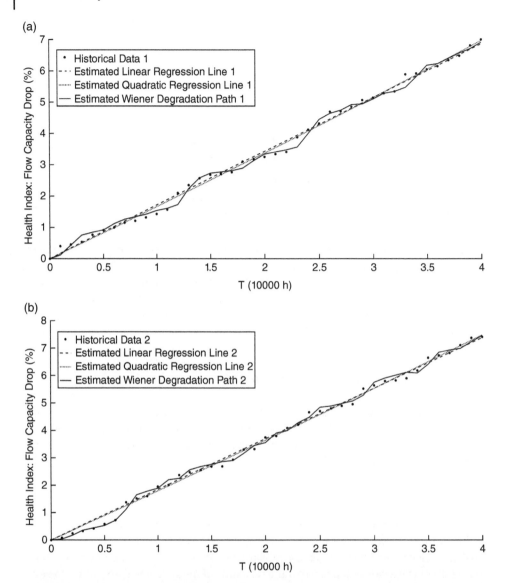

Figure 2.4 Comparisons among the Wiener degradation model of (2.3) and linear and quadratic degradation models used in [161, 288] for two gas turbines.

where

$$\varphi(x) = \begin{cases} 1, & if \quad x \neq 1 \\ 0, & else \end{cases},$$

$$\theta = 2 \cdot RT_{max} + PO_{max} + ST_{max} \text{ or } 2 \cdot DT_{max} + PO_{max} + ST_{max}$$

and

$$Q = \sum_{h=1}^{H} \left(\frac{\sum_{i=1}^{N_T} (2 \cdot RT_i + PO_i + ST_i)}{N_T \cdot H} \right) \text{ or } \sum_{h=1}^{H} \left(\frac{\sum_{i=1}^{N_T} (2 \cdot DT_i + PO_i + ST_i)}{N_T \cdot H} \right)$$

During each PSO iteration, information about the best position of each particle and the best particle in the entire swarm can be obtained after computing the fitness values of all particles based on (2.16). And then in the next iteration, the movements of the particles are guided by their previous best-known positions within the search space, called *pbest*, as well as the best known position of the entire swarm, called *gbest*.

For a particle j with D dimensions, its local best known position can be represented by $pbest_j = (p_{j1}, p_{j2}, ..., p_{jD})$ and the global best known position of the entire swarm is $gbest = p_g$. Therefore, the velocity v_{jd}^{m+1}, and the position x_{jd}^{m+1}, associated with the dimension d of the particle j after iteration m are updated according to:

$$v_{jd}^{m+1} = w \cdot v_{jd}^m + c_1 \cdot \rho_1 \cdot (p_{jd} - x_{jd}^m) + c_2 \cdot \rho_2 \cdot (p_g - x_{jd}^m) \tag{2.17}$$

$$x_{jd}^{m+1} = x_{jd}^m + v_{jd}^m \tag{2.18}$$

where $m \in \{1, ..., M\}$; $j \in \{1, ..., J\}$; $d \in \{1, ..., D\}$; w is the inertia weight; c_1 and c_2 are the cognition learning factor and the social learning factor; $c_1 + c_2$ generally equals to 4 and ρ_1 and ρ_2 are random factors restricted to $[0, 1]$.

The followings are some important details of the proposed MCS-PSO method:

STEP 1: Randomly initialize the particles representing the PID control strategies, including the velocity and the position.

STEP 2: Apply the MCS proposed in section 2.2.2.2 to obtain the $R^*(T_o)$; then compute the fitness value for each particle based on (2.16).

STEP 3: Update the values of *pbest* and *gbest* and determine the velocity and position values for each particle in the next iteration on the basis of (2.17) and (2.18).

STEP 4: If the stopping condition - the maximum number of iterations M is reached, go to **STEP 5**; otherwise, go to **STEP 2**.

STEP 5: Decode the particle with the minimum fitness value and take the result as the optimally designed PID control strategy which can ensure the maximum possible reliability or the best quality parameters associated with the control process with reliability at 1 at total run time T_o.

2.3.2 Optimal PID-AGC

The AGC performance of the integrated system depends on the discrete-time PID controller. Therefore, the PID controller is optimized to mitigate the communication network disturbances and offer optimum AGC performance by reducing system frequency fluctuations. As a result, system reliability is also optimized. In previous works, the objective function for the optimization of the PID controller is an integral performance index over the total operating time T, which quantifies frequency and control signal deviations [79, 289]:

$$J = \int_0^T [\eta_1 (\Delta f(t))^2 + (1 - \eta_1) * \eta_2 (\Delta u(t))^2] dt \tag{2.19}$$

where η_1 indicates the relative importance of the two terms and η_2 is the normalizing constant to scale both terms in a uniform range and is set to 0.002. The first term is directly related to the reliability of the integrated system and the second term measures the disturbance rejection ability of the controller, which is the total control effort to be minimized. The total process time T is set to 30 s [289].

Motivated by [271–275, 290], the objective function in (2.19) is separated into two objectives and multi-objective optimization is applied. In fact, the weighting factors in the objective function in (2.19) can be changed dynamically using multi-objective PSO algorithm:

$$\text{Minimize: } J(\vec{x}) = [\text{E}[J_1(\vec{x})], \text{E}[J_2(\vec{x})]]^T \tag{2.20}$$

where $J_1(\vec{x}) = (\Delta f(t))^2$ and $J_2(\vec{x}) = (\Delta u(t))^2$. $J(\vec{x}) : \mathcal{R}^3 \to \mathcal{R}$ is defined in (2.20). The variable $\vec{x}$ are the controller parameters, i.e., K_P, T_I and T_D, and the 3-dimensional search space G $\in$ $\mathcal{R}^3$ is pre-specified as G $= [0, 10]^3$ to widely span the optimization range of the controller design [79].

The fitness value of the stochastic population-based algorithms is the expectation of the stochastic objective function obtained from MCS:

$$\text{E}[J_i] = \sum_{j=1}^{N} J_{i,j}/N, i = 1, 2 \tag{2.21}$$

where J_i is defined in (2.20) and $N = 200$ samples.

In this section, we use the PSO rather than other heuristic algorithms, i.e., GA [74], to solve the optimization problem for $\vec{x} \in \mathcal{R}^3$, because PSO can avoid being trapped into local minima [78–80]. The particle movement in the PSO can be interpreted as a form of path re-linking via the exploitation of the local best-known positions, i.e., pbest [96].

In this sense, both the PSO and GA can be regarded as generating new solutions in the neighborhood of two parents, i.e., using crossover in the GA and using attractions to two pbest positions in the PSO. This multi-parent effect is a key advantage over single-point techniques such as the simulated annealing and tabu search.

As compared to GA, the PSO utilizes three available information sets to facilitate the search process. These are the local best-known positions (pbest), the global best-known position (gbest), and the current positions [78–80]. This allows greater diversity and exploration over a single population (which with elitism would only be a population of pbests).

In addition, the momentum effects on particle movement allows fast convergence (e.g., when a particle is moving in the direction of a gradient) and increased variety/diversity in search trajectories. The performance of PSO has been shown superior to GA in parameters optimization of PID controller [291, 292]. The MOPSO is preferred to the multi-objective GA because no evolution operators, i.e., the crossover and mutation, are required, and the information of the current optimum particles effectively identifies candidate solutions throughout the problem space.

The detailed introduction to PSO and MOPSO are provided in [78–80] and [271–275, 290], respectively. The core part of the MOPSO algorithm is detailed in the following.

Notation to be used in MOPSO Algorithm:

A: solution set which provides optimal values of the PID controller parameters K_P, K_I and K_D

NP: number of particles

x_i and v_i: current position and velocity of particle i

MP: number of dimensions ($MP = 3$)

I and NI: current and maximum number of iterations

ω: inertia coefficient ($\omega = 0.5 \times 0.99^I$ which tunes the impact of the previous history velocities on the current velocity)

c_1 and c_2: cognitive and social factors (which can accelerate the search towards the local and global best directions, and equal to $c_1 = 1$ and $c_2 = 2$)

r_1 and r_2: random numbers drawn from the uniform distribution $[0, 1]$.

Procedures of MOPSO Algorithm:

STEP 1: Randomly initialize the position x_i of the particle i ($i = 1, 2, ..., NP$) sampling from a uniform distribution within the solution space, i.e., $G = [0, 10]^3$. The coordinates of the particle position represent the PID controller parameters, i.e., K_P, K_I and K_D. Initialize v_i, $pbest_i$ and $gbest_i$ to zero.

STEP 2: Compute the fitness value for each particle based on (2.20).

STEP 3: Update the values of *pbest* and *gbest* based on calculated fitness values; determine the velocity and position values for each particle in the next iteration using following equations:

$$v_{id} := \omega v_{id} + c_1 r_1 (P_{id} - x_{id}) + c_2 r_2 (G_{id} - x_{id}) \tag{2.22}$$

$$x_{id} := x_{id} + v_{id} \tag{2.23}$$

where $P_i = [pbest_{i1}, pbest_{i2}, ..., pbest_{iMP}]^T$, $G_i = [gbest_{i1}, gbest_{i2}, ..., gbest_{iMP}]^T$ and $d = 1$, 2, ..., MP. The current best solution is put into the set A.

STEP 4: If the maximum number of iterations NI is reached, go to **STEP 5**; otherwise, go to **STEP 2**.

STEP 5: Decode the particle with the minimum fitness value and choose the result as the optimally designed PID controller.

As indicated by [271–275, 290], the multi-objective optimization generates the Pareto optimal set of the non-dominated solutions from which the global optimum solution is selected based on the specific application. In this study, fuzzy set theory is applied to model the trade-off between the two objectives [271–275, 290]. First, a linear membership function is defined for each objective function $J_i(\vec{x})$:

$$mf_i = \frac{J_i - J_i^{min}}{J_i^{max} - J_i^{min}}, i = 1, 2 \tag{2.24}$$

According to (2.24), the minimum of the objective function J_i is associated to the minimum of the membership function, therefore it has maximum degree of achievement of the fuzzy objective. Then, for every non-dominated solution k, the aggregate membership function for the PID controller optimization is computed as:

$$MF1^k = mf_1^k + mf_2^k \tag{2.25}$$

The solution which minimizes (2.25) is regarded as the best compromise between the two objectives and named "Best Tradeoff I". For comparison, the distance of each non-dominated solution k from the origin in the fuzzified coordinate is also computed [271–275, 290]:

$$MF2^k = [(mf_1^k)^2 + (mf_2^k)^2]^{1/2} \tag{2.26}$$

And the solution with minimum $MF2^k$ is chosen as the best compromise between the two objectives and named "Best Tradeoff II". The convergence of optimization problem solved by the single PSO has been investigated in detail in works [79, 283]. These works investigated similar optimization problem of PID controller, where both the objective value and the controller parameters become almost constant toward the end of 50 iterations (the number of populations is 30). As for the convergence of optimization problem of PID controller solved by the MOPSO, works [271–275, 290] showed fast convergence to the Pareto front in 50 iterations (the number of populations is 50).

3

Stability Enhancement of CPSs

In this chapter, we preform the stability analysis and enhancement method on one of the most widely used Cyber-Physical Systems (CPSs) - microgrids from two perspectives as illustrated example: 1) the operations of the integrated Distributed Energy Resource (DER) systems and open communication network, and 2) the design of optimal Automatic Generation Control (AGC) strategies in the face of communication degradation.

This chapter focuses on microgrids in islanded operations with real power generation and demand from a systemic perspective. The total generation of DERs supplies the demand. Therefore, the initial conditions of DERs are determined to balance the stochastic Renewable Energy Resources (RERs) and the demand. As a result, voltage profiles and reactive power are neglected. This simplified model is adopted by several works [74, 76, 79], and is also integrated with microgrid power management systems [97–102].

Because wind and solar sources have intermittent characteristic, their power generations have stochastic behaviors. The probabilistic power patterns can be well justified by probability distribution functions, such as Weibull and normal distribution for wind speed and solar irradiation [273, 295]. Alternatively, in this chapter we consider actual wind speed and solar irradiation as in [296, 297]. The historical wind speed and solar radiation dataset are from the Elia Grid, Belgium.

The ability of the integrated system to maintain system frequency deviation within tolerance margins quantifies system reliability and is evaluated by Monte Carlo simulation (MCS) as detailed in Section 2.2.1.2. Stochastic time delays are modelled by generating random congestion based on Media Access Control (MAC) protocols. Congestion of network channels depends on the activity level of the interfering traffic, which is the root cause of network-induced delays and packet dropouts [78, 88, 94, 298]. The open communication network model is implemented via TrueTime simulator testing different MAC protocols [92, 283] as detailed in Section 2.1.2.

To stabilize system frequency against RER, demand variability and communication degradation, a discrete Proportional–Integral–Derivative (PID) controller is used [79, 289]. The heuristic algorithms such as PSO [78–80, 96, 299], Genetic Algorithms (GA) [74], flower pollination algorithm [75], quasi-oppositional harmony search algorithm [76], Cuckoo Search algorithm [87], and artificial bee colony algorithm [300], have been introduced to provide easy implementation, conceptual simplicity, more flexibility, and independency to the initialization of the controller optimization procedure.

Part of the contents are from [293, 309] and permission has been obtained to use the contents.

Cyber-Physical Distributed Systems: Modeling, Reliability Analysis and Applications, First Edition.
Huadong Mo, Giovanni Sansavini and Min Xie.
© 2021 John Wiley & Sons Ltd. Published 2021 by John Wiley & Sons Ltd.

PSO is adopted to minimize the stochastic objective function and achieve the optimal PID controller for various architectures and conditions of the open communication network, because the movement of particles is influenced by the local best-known positions as well as by the global best-known positions in the search space, PSO can avoid being trapped into local minimum [78–80, 96, 299]. Finally, the robustness and effectiveness of the proposed MOPSO-based-PID-controlled AGC against communication degradation is assessed.

In addition, we conduct various simulations to test the performance of the integrated system with the MOPSO-optimized control strategy for three communication network architectures, under different uncertainty conditions and operating scenarios of the DERs and load. The performance of the integrated system with the perfect communication is used as a benchmark.

The other two network architectures are the Ethernet and the hybrid network (a mix of Ethernet and 802.11b/g). Different network configurations, i.e., traffic conditions of the open communication networks, are assessed. The performance of the integrated system using the three network architectures are compared under the uncertainty conditions and the effectiveness and robustness of the proposed control strategy is shown.

3.1 Integration of Physical and Cyber Models

3.1.1 Basics of WAPS

3.1.1.1 Physical Layer

Figure 3.1 shows the control block diagram of a single-area Wide-Area Power System (WAPS) in which one generator supplies power to the service area and the transmissions of frequency measurements and control signals occur via the open communication network [23, 24]. It is extended from the diagram presented in Figure 2.1.

The WAPS physical configuration is provided in Figure 3.1. ΔP_v, ΔP_m, ΔP_d, u_k and Δf_k are the valve position, the mechanical output of the turbine, the load, the control signal computed by the controller in the period k and the deviations of frequency measured by the Phasor Measurement Units (PMU) in the period k, respectively. The droop characteristic is a feedback gain to improve the damping characteristics of the power system and is usually set to $1/R$.

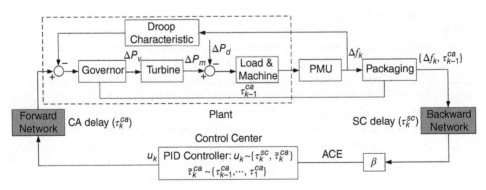

Figure 3.1 Schematic of the single-area WAPS with communication delays.

The dynamics of the governor, the non-reheat turbine, and the load and machine are described by the first order transfer functions with specified gain and time constant as:

$$G_G(s) = \frac{1}{T_G s + 1} = \frac{\Delta P_v}{u_k - \frac{\Delta f_k}{R}} \tag{3.1}$$

$$G_T(s) = \frac{1}{T_T s + 1} = \frac{\Delta P_m}{\Delta P_v} \tag{3.2}$$

$$G_P(s) = \frac{1}{Ms + D} = \frac{\Delta f_k}{\Delta P_m - \Delta P_d} \tag{3.3}$$

where T_G, R, T_T, M and D denote the time constant of the governor, speed drop, time constant of the turbine, moment of inertia of the generator and generator damping coefficient.

As there is no power exchange to other areas, the area control error (ACE) in period k is defined as:

$$ACE = \beta \Delta f_k \tag{3.4}$$

where $\beta > 0$ denote the frequency bias factor.

The PID controller is tuned to improve the dynamic performance of Load Frequency Control (LFC) strategy by balancing the generation production and the demand load in real time. It receives real-time system frequency measurements from the PMU via the open communication network and then computes the control signals based on (3.5).

If Δf_k is negative, the PID controller will send out a command to increase the generation production; and if Δf_k is positive, the PID controller is going to reduce the generation production. However, the PMU takes measurements at every sampling time, therefore the PID controller is time-discrete rather than time-continuous as in [18]. Its transfer function $C(z)$ is:

$$C(z) = K_p \left(1 + \frac{a(z)}{T_i} + T_d \cdot \frac{N}{1 + Nb(z)} \right) \tag{3.5}$$

where K_p, T_i and T_d are proportional, integral and derivative gain of the PID controller. $a(z)$ and $b(z)$ are determined by the integrator method and filter method. N is the filter constant. The forward Euler method is most appropriate for the case with a small sampling interval, where the Nyquist limit is large compared to the bandwidth of the controller.

As the sampling interval of PMU is rather small, the integrator method and filter method in this chapter apply the forward Euler method, where $a(z)$ and $b(z)$ are all equal to $\frac{T_s}{z-1}$ (T_s is the sampling interval of the PMU).

3.1.1.2 Cyber Layer
The cyber layer in WAPS is mainly based on the open communication network, where the total time delay can be expressed as [301]:

$$T_{tot} = T_{pre} + T_{wait} + T_{tx} + T_{post} \tag{3.6}$$

where T_{pre} denotes the preprocessing time at the source which is the sum of the computation time T_{scomp} and the encoding time T_{scode}. T_{wait} denotes the waiting time at the source which is the sum of the queue time T_{queue} and blocking time T_{block}. T_{tx} denotes the network time delay which is the sum of the time required to send the data to the channel T_{frame} and the propagation time T_{prop}. T_{post} denotes the postprocessing time at the destination, which is the sum of the decoding time T_{dcode} and the computation time T_{dcomp}.

T_{pre}, T_{tx} and T_{post} depend heavily on the processing speed of software or firmware of devices, the packet size and the length of the network cable, which are typically constant and neglectable.

For example, to send one bit of data to a 10 Mb/s Ethernet requires $T_{frame} = 67.2$ μs and $T_{prop} = 10$ μs for a 2500 m Ethernet with typical transmission rate 2×10^8 m/s. Due to the uncertain amount of data to be sent and the stochastic traffic on the network channel, T_{wait} is the major source of jitter and usually determined by the MAC protocol [57].

This section considers the Ethernet with Carrier Sense Multiple Access with Collision Detection (CSMA/CD) based MAC protocol and then T_{wait} can be analyzed. T_{queue} is the time that a packet waits in the buffer of the source node while previous packets in the queue are sent, depending on the blocking of previous packet in queue, the periodicity of packets and the processing load. As T_{block} plays a critical role in T_{queue}, the work mainly investigates T_{block}.

T_{block} in the Ethernet is consisted of the time taken by collisions with other packets and the subsequent retransmitted time. As T_{block} is probabilistic indicated by the BEB algorithm, it is very difficult to derive an exact expression for T_{block} and thereby most current studies focus more on its expectation. When the k-th collision happens, the sender at source node would back off for a time T^{bo} defined by

$$T_k^{bo} = \frac{minimum\ frame\ size}{data\ rate} \cdot RN \tag{3.7}$$

where RN follows a discrete uniform distribution taking value from $(0, 2^k - 1)$. For the Ethernet with CSMA/CD MAC protocol, the maximum number of k is 10 [283].

Therefore, the expected T_{block} can be expressed by the following equation as

$$E\{T_{block}\} = \sum_{k=1}^{10} E\{T_k^{bo}\} + T_{resid} \tag{3.8}$$

where $E\{T_k^{bo}\}$ is the expected back off time of the k-th collision and T_{resid} is the residual time that is seen by the source until the network is idle. For the 10-th collision, the source would discard this packet and report an error message to the higher-level processing units. As $E\{T_k^{bo}\}$ depends heavily on the number of backlogged and unbacklogged nodes as well as arrival rate at each node in the network, T_{block} is not deterministic and cannot be bounded due to the discarding of packet [283].

3.1.1.3 WAPS Realized in TrueTime

The integrated model of the single-area WAPS and the CSMA/CD MAC protocol-based Ethernet is implemented via the TrueTime [283], and the data rate is $8 \cdot 10^4$ bits/s. The architecture of the integrated system and its process overview during the control period k, as shown in Figure 3.1, are listed as follows. It is more detailed than the general framework presented in Section 2.1.2.2.

- The PMU node (periodic): based on the NREL report [302] and the standard *C37.118.1* [303], the synchronized PMU is assumed to sample AC waveforms with a sampling interval T_s of 0.3 second. The packet (average size = 400 bits [53, 58]), containing the time delay between the control center to the plant τ_{k-1}^{ca} in the previous $k - 1$ period and the calculated frequency Δf_k, is labelled with a GPS time stamp t_1, and then sent to the control center over the Ethernet. The delay between the sampling time and the queuing time is $5 \cdot 10^{-4}$ s [58].

- The control center node (event-driven): when the control center receives a frequency measurement Δf_k from the PMU, the delivering packet is re-labelled with a GPS time stamp t_2; thus, the time delay between the PMU signal and the control center is $\tau_k^{sc} = t_2 - t_1$. The controller calculates the control signal u_k based on (3.5), which also fits a packet of 400 bits [53, 58] and sends it to the plant. This packet is labelled with a GPS time stamp t_3. The time for computing the control signal is set to $5 \cdot 10^{-4}$ s [58].
- The plant node (event-driven): the plant acts based on the control signal received, which is re-labelled with a GPS time stamp t_4. The time delay in the communication from the control center to the plant is $\tau_k^{ca} = t_4 - t_3$. The delay between the reception of the packet and the actuation is $5 \cdot 10^{-4}$ s [58].
- The interfering node (periodic): it sends out the disturbing traffic over the network to mimic congestion and generate time delays. This node sends out packets (size $\in [80, 400]$ bits [58]) with a probability indicating the fraction of the network bandwidth, i.e., *BWshare*, occupied by other users in the open communication network. The period of interfering trigger T_{IT} denotes the number of active end-users.

The time delays can be obtained from the GPS-based time stamp technique and the network-induced time delays in C-A channel are displayed in Figure 3.2. These delays depend heavily on the network state, which reflects the severity of network load and congestion levels [53]. The network state is categorized into 'good', 'normal' and 'bad'.

In period k, the controller cannot have any knowledge of the future τ_{k+1}^{sc} and τ_k^{ca}. Therefore, it is necessary to predict the τ_k^{ca} and τ_{k+1}^{sc} before the controller computes the control signal to dispatch. The predicted S-C time delay, $\hat{\tau}_{k+1}^{sc}$, and the predicted C-A time delay, $\hat{\tau}_k^{ca}$, can be effectively compensated by the Smith predictor or by a delay-dependent controller.

3.1.2 An Illustrative WAPS

The schematic of the illustrative WAPS is detailed in this section, which is made of a physical layer as shown by Figure 3.3, a cyber layer as shown by Figure 3.5 and resulting integrated system as shown by Figure 3.6.

3.1.2.1 Illustrative Physical Layer

The structure of physical layer in Figure 3.3 is general, representative of the Distributed Generation Systems (DGS), and widely adopted in the literature [74, 76, 79, 290]. It models

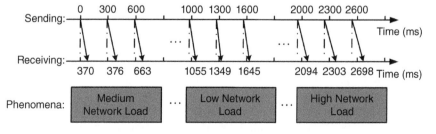

Figure 3.2 An illustrative example for the time delays in C-A channel.

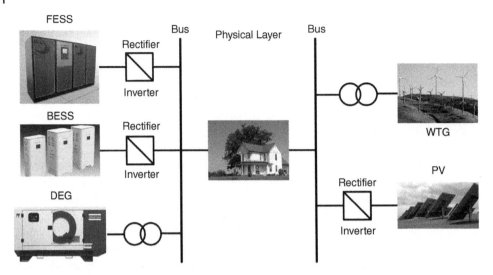

Figure 3.3 Schematic of illustrative physical.

a hybrid microgrid, which consists of conventional generators (Diesel Engine Generator (DEG)), RERs (wind turbine generator, WTG, and photovoltaic power, PV) and ESSs (Battery Energy Storage System (BESS) and Flywheel Energy Storage System (FESS)).

The small signal stability analysis of the hybrid microgrid in Figure 3.5, is based on time-domain simulations, i.e., transfer function models. In the AGC, the WTG, PV, DEG, FESS and BESS are described by first order transfer functions with specified gain and time constant. A centralized controller is used [76], as opposed to multiple decentralized controllers for each controllable component [74, 304, 305]. It enables easier maintenance and reduces wiring cost and makes the AGC design problem traceable by reducing the number of controller parameters [78].

On the other hand, the centralized controller impacts the AGC performance, because a unique control signal is used by all the components. Nevertheless, current studies show that the centralized controller can ensure acceptable time-domain AGC performance [76, 79, 96].

The transfer functions $G_{WTG}(s)$, $G_{PV}(s)$, and $G_{DEG}(s)$ of the WTG, PV and DEG, respectively, are expressed as

$$G_{WTG}(s) = \frac{1}{1 + sT_{WTG}} = \frac{P_{WTG}}{P_W} \tag{3.9}$$

$$G_{PV}(s) = \frac{1}{1 + sT_{PV}} = \frac{P_{PV}}{P_{sol}} \tag{3.10}$$

$$G_{DEG}(s) = \frac{1}{1 + sT_{DEG}} = \frac{P_{DEG}}{\widetilde{u_{DEG}}(t)} \tag{3.11}$$

where T_{WTG}, T_{PV} and T_{DEG} are the time constant of the WTG, PV and DEG. P_{WTG} and P_{PV} are the electrical power produced from the RERs.

Wind power P_W and solar power P_{sol}, which are generated from the data sets of wind speed and solar radiation of the Elia Grid, Belgium, and their curves are provided in the Figure 3.4 (a), Figure 3.4 (b) and Figure 3.4 (c), respectively.

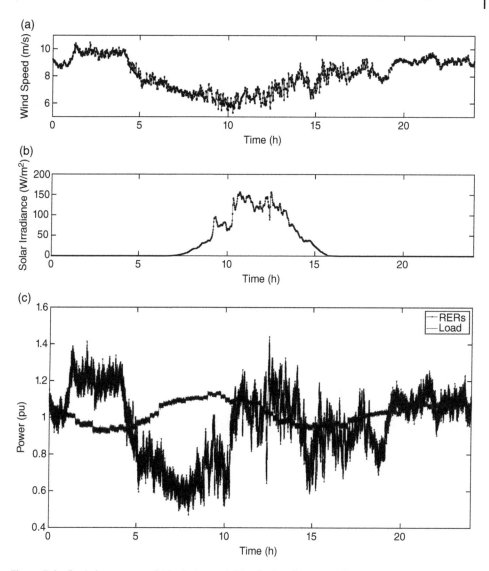

Figure 3.4 Real-time curves of (a) wind speed, (b) solar irradiance and (b) comparison between the total output power of RERs and the load.

The DEG is controlled by the control signal $\widetilde{u_{DEG}}(t)$ sent by the remote-control center and it generates power only when the RERs cannot meet the demand.

The ESSs are critical in eliminating frequency fluctuations due to their fast response to the control signal. Based on [77, 79, 290, 305], the transfer functions $G_{FESS}(s)$ and $G_{BESS}(s)$ of the FESS and BESS are given as

$$G_{FESS}(s) = \frac{1}{1 + sT_{FESS}} = \frac{P_{FESS}}{\widetilde{u_{FESS}}(t)} \tag{3.12}$$

$$G_{BESS}(s) = \frac{1}{1 + sT_{BESS}} = \frac{P_{BESS}}{\widetilde{u_{BESS}}(t)} \tag{3.13}$$

where T_{FESS} and T_{BESS} are the time constant, P_{FESS} and P_{BESS} are the output power, $\widetilde{u_{FESS}}(t)$ and $\widetilde{u_{BESS}}(t)$ are the control signal of the FESS and BESS.

Remark 3.1 The DEG, FESS and BESS have rate constraint, i.e., $|P_{DEG}| < \overline{P}_{DEG}$, $|P_{FESS}| < \overline{P}_{FESS}$ and $|P_{BESS}| < \overline{P}_{BESS}$, where $\overline{P}_{DEG}$, $\overline{P}_{FESS}$ and $\overline{P}_{BESS}$ are the maximum rated output power of the DEG, FESS and BESS, respectively.

The transfer function $G_{HPS}(s)$ of the hybrid power system (HPS) models the relationship between the power imbalance, i.e., $\Delta P_S - \Delta P_L$, and the system frequency $\Delta f(t)$

$$G_{HPS}(s) = \frac{1}{D + Ms} = \frac{\Delta f(t)}{\Delta P_S - \Delta P_L} \tag{3.14}$$

where M and D are the inertia constant and damping constant of the HPS [78], and P_S is the total power generated, denoted by $P_{WTG} + P_{PV} + P_{DEG} + P_{FESS} + P_{BESS}$. P_L is the power demand and its curve is provided in Figure 3.4 (c).

3.1.2.2 Illustrative Cyber Layer

The cyber layer Figure 3.5 is responsible for the real-time communication between the control center and subsystems, i.e., providing data exchanges between the control center and controllable BESS, FESS and DEG in the physical layer. In addition, PMU measurements and control signals are transmitted via the shared and open communication network.

The time delays τ_{sc} and τ_{ca} are determined by (3.6), and their impact on the transfer function from the control signal $U(s)$ to the system output $Y(s)$ is described as

$$\frac{Y(s)}{U(s)} = G(s)e^{-(\tau_{sc} + \tau_{ca})s} \tag{3.15}$$

where $G(s)$ denotes the transfer functions of the DERs.

A second source of disturbance in the source-destination communication is packet dropout, which occurs for three major reasons, i.e., the network disconnection, time-out transmission and time-out re-transmission [283]. The PMU sensor is time

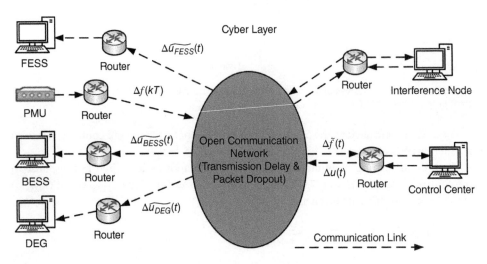

Figure 3.5 Schematic of illustrative cyber layer.

triggered, i.e., it takes measurement at every sampling interval T_s. The relationship between the frequency measured by the PMU and the frequency received by the control center is

$$\Delta \widetilde{f}(t_n) = \gamma_k \Delta f(kT_s) \tag{3.16}$$

where $\gamma_k = 1$ indicates the transmission of $\Delta f(kT_s)$ is successful and t_n is the time instant when the n-th data packet is received by the control center. If $\gamma_k = 1$, $\tau_{sc}^n = (t_n - kT_s)$, where τ_{sc}^n is the transmission time of packet n from the PMU to the control center.

In real implementations of the CSMA/CD MAC protocol, the source node discards the packet and reports an error if ten consecutive data collisions occur [283]. Therefore, if congestion is severe, time delays increase and so does the number of packet dropouts. However, network congestion and packet dropout are challenging to link, and, therefore, packet dropout is usually modelled as a stochastic process [282, 283, 307]. Binary switching sequences are usually applied [105], which specify the expected packet dropout probability.

The stochastic parameter γ_k of the binary switching sequence is Bernoulli distributed [106, 307], taking value of 0 or 1 with

$$P\{\gamma_k = 0\} = L_d \tag{3.17}$$

where $0 < L_d < 1$ is the expected packet loss probability.

The frequency measurement $\Delta \widetilde{f}(t)$ is not updated until the next packet is received, thus the evolution of $\Delta \widetilde{f}(t)$ is given as

$$\Delta \widetilde{f}(t) = \Delta \widetilde{f}(t_n), \quad t_n \leq t < t_{n+1} \tag{3.18}$$

On the other hand, the controller is event-driven, i.e., it updates the control signal and sends it to the DERs as soon as the control center receives an updated frequency measurement. Therefore, if the n-th packet is received, the control signal $\Delta \widetilde{u}_i(t)$, stored in the buffer of the DER i (i=FESS, BESS, DEG) is updated as

$$\Delta \widetilde{u}_i(t) = \Delta \widetilde{u}_i(t_m), \quad t_m \leq t < t_{m+1} \tag{3.19}$$

$$\Delta \widetilde{u}_i(t_m) = \gamma_n u(t_n) \tag{3.20}$$

where $\gamma_n = 1$ if the control signal $u(t_n)$ computed based on $\Delta \widetilde{f}(t_n)$ is not dropped (and $\gamma_n = 0$ otherwise) and t_m is the time at which the m-th data packet is received by the DER i. If $\gamma_n = 1$, $\tau_{ca}^m = (t_m - t_n)$, where τ_{ca}^m is the transmission time packet m from the control center to the DER i. The control center sends control signals to the FESS first, then to the BESS and finally to the DEG.

3.1.2.3 Illustrative Integrated System

The cyber layer and the physical layers have the same topology. Power imbalance is the root cause of system frequency fluctuations. The control center remotely monitors the Energy Storage System (ESS) and the diesel generator to reduce the imbalance and to ensure good AGC performance. Following [79, 290], the PID controller in the control center is responsible for the system frequency stabilization, and therefore we use it to represent the control center in the control block diagram shown in Figure 3.6.

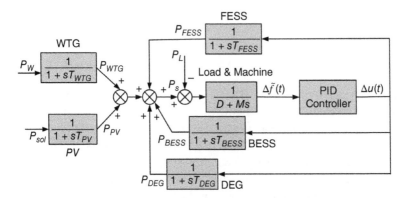

Figure 3.6 Control block diagram of illustrative integrated system.

According to work [285], the linearized state-space realization of the microgrid in the physical layer of Figure 3.6 can be expressed as

$$\dot{x} = Ax + B_1 w + B_2 u$$

$$y = Cx \tag{3.21}$$

where

$$x^T = [\Delta P_{WTG} \Delta P_{PV} \Delta P_{DEG} \Delta P_{BESS} \Delta P_{FESS} \Delta f] \tag{3.22}$$

$$w^T = [\Delta P_W \Delta P_{sol} \Delta P_L] \tag{3.23}$$

$$y = \Delta f \tag{3.24}$$

An elaborated expression of the microgrid state-space model is given as follows.

$$\dot{x} = \begin{bmatrix} -1/T_{WTG} & 0 & 0 & 0 & 0 & 0 \\ 0 & -1/T_{PV} & 0 & 0 & 0 & 0 \\ 0 & 0 & -1/T_{DEG} & 0 & 0 & 0 \\ 0 & 0 & 0 & -1/T_{BESS} & 0 & 0 \\ 0 & 0 & 0 & 0 & -1/T_{FESS} & 0 \\ 0 & 0 & 0 & 0 & 0 & -2D/M \end{bmatrix} \begin{bmatrix} \Delta P_{WTG} \\ \Delta P_{PV} \\ \Delta P_{DEG} \\ \Delta P_{BESS} \\ \Delta P_{FESS} \\ \Delta f \end{bmatrix}$$

$$+ \begin{bmatrix} 1/T_{WTG} & 0 & 0 \\ 0 & 1/T_{PV} & 0 \\ 0 & 0 & 0 \\ 0 & 0 & 0 \\ 0 & 0 & 0 \\ 0 & 0 & 2/M \end{bmatrix} \begin{bmatrix} \Delta P_W \\ \Delta P_{sol} \\ \Delta P_L \end{bmatrix} + \begin{bmatrix} 0 \\ 0 \\ 1/T_{DEG} \\ 1/T_{BESS} \\ 1/T_{FESS} \\ 0 \end{bmatrix} u \tag{3.25}$$

$$y = [0\,0\,0\,0\,0\,1]x \tag{3.26}$$

3.2 Settings of Stability Analysis

In this chapter, case studies are carried out on the LFC strategy of the illustrative single-area WAPS with communication delays and are compared with results of the delay margin-based method [17] and the evolution algorithm-based method [35]. This section introduces the simulation settings for time delay predictions and different cases used in the LFC of DERs.

3.2.1 Settings for Delay Predictions

The parameters of the integrated system are in Table 3.1 [17, 44, 57], where T_s is the sampling interval of the PMU, *BWshare* is the threshold used in the interfering node and T_{IT} is the period of the interference trigger. This table is specifically used in the time delay predictions in the Section 3.3.

The standard *C37.118.1* is applied for the synchronized PMU in power systems, in a single-area WAPS with one PMU and two phasors [282]. The data frame of PMU measurement is of the order of few hundreds of bytes and the transmitted data rate is 80000 bits/s. The time-tagged phasors are sent to the control center at rates up to 60 samples per second. Due to the small packet size of PMU measurement, single-packet transmission is considered, i.e., the data are lumped into one packet. Single-packet transmission is adopted by the MAC protocols with large packet size, e.g., the maximum transmission unit in the Ethernet is 1500 bytes and the maximum transmission unit in the Wireless Local Area Networks (WLAN) is 7981 bytes. In this chapter, the open communication network employs the Ethernet.

The results demonstrate the accuracy of the proposed Discrete Hidden Markov Model (DHMM) method in predicting time delays (Section 3.3.2). Simulation studies investigate the performance of LFC of the single-area WAPS, which is equipped with a Proportional–Integral (PI) controller tuned via the delay margin method (**Case 1** in Section 3.3.3.2) and a PID controller tuned via evolutionary algorithms (**Case 2** in Section 3.3.3.3), with and without the Smith predictor. This comparison demonstrates the effectiveness of the DHMM-based Smith predictor in improving the LFC performance and reducing frequency oscillations.

3.2.2 Settings for Illustrative WAPS

In the illustrated WAPS to be used for the LFC of RERs, the physical system operates in nominal conditions, i.e., the stochastic wind speed, the variable sun irradiance and

Table 3.1 Parameters of the single-area WAPS with communication network.

Parameter	T_G	T_T	M	D	R	β
WAPS	0.1	0.3	10	1	0.05	21

Parameter	T_s	BWshare	T_{IT}	μ_P	σ_P	K_C
Network	0.3	0.2	0.002	130	55	100

uncertain load, respectively, P_{WTG}, P_{PV} and P_L, given in [74, 79, 100]. The coupled algebraic and ordinary differential equations for the DGS and the open communication network in Figure 3.3, are numerically integrated using the Dormand-Prince method implemented in Matlab ode45 function with a fixed step size of 0.005 s. The parameters of the transfer functions for the DERs in Figure 3.3 are provided in [74, 79, 100]. Additionally, the physical layer has the following specifications:

- The base value for the apparent power is 150 kW.
- The rated apparent power of the WTG (Nordtank NTK 150) is 150 kW. The cut-in wind speed, the rated wind speed and the cut-out wind speed are 4.0 m/s, 13.5 m/s and 25 m/s, respectively. The wind farm consists of two WTGs.
- The structure of a 750 kW PV generation has 465 parallel strings and consists of 7 series-connected modules (SunPower SPR-230E-WHT-D). The rated power of each cell is 230 W under the nominal operation temperature 45 °C. The nominal efficiency is 18.5%. The length and width of each cell are 1.599 m and 0.798 m. The maximum power temperature coefficient is 3.37×10^{-5}.
- The rated power of the DEG (MQP300IV) is 300 kW.
- The rated power of the BESS (one Tesla Powerpacks) is 50 kW, whose energy capacity is 210 kWh.
- The rated power of the FESS (Amber Kinetics) is 25 kW, whose energy capacity is 40 kWh.

Based on *Remark 3.1*, $0 \leq \Delta P_{DEG} \leq 2$ pu, $|\Delta P_{BESS}| \leq 0.33$ pu and $|\Delta P_{FESS}| \leq 0.17$ pu, provide the largest rated output of the DEG, FESS and BESS [79]. The data sets of wind speed and solar radiation for computing real-time output of RERs are provided by the Elia Grid, Belgium, with a data frequency 1 Hz. The details can be referred to Figure 3.4. The relevant system parameters [79, 290] for a typical microgrid in the physical layer of Figure 3.3 are presented in Table 3.2.

The open communication network is implemented in TrueTime simulator [283]. The detailed procedures are provided in the Section 2.1.2. The data rate of the communication architectures is 800 Kbits/s, and the 802.11b/g is further characterized by transmission power 20 dbm, receiver signal threshold - 48 dbm, ACK timeout 0.04 ms and the retry limit 3. The descriptions of simulation parameters T_i, BWShare and L_d is discussed in *Remark 2.1*, *Remark 2.2* and (3.17) and presented in Table 3.3.

3.2.3 Cases for Illustrative WAPS

In order to simulate diverse operating conditions of the islanded microgrid, four exemplary cases of the DERs management are considered. The power generated by the RERs and the

Table 3.2 Parameters of frequency response model in the Figure 3.3.

Parameter	Value	Parameter	Value
D (pu/Hz)	0.03	T_{DEG} (s)	2
M (pu/s)	0.4	T_{WTG} (s)	1.5
T_{BESS} (s)	0.1	T_{PV} (s)	1.8
T_{FESS} (s)	0.1		

Table 3.3 Parameters of communication network in Figure 3.5.

Parameter	Value	Descriptions of Value
BWShare	$0\sim1$	indicates the expected percentage of bandwidth used by the interference user
T_i	$0\sim+\infty$	determines the activity level of the interference user
L_d	$0\sim1$	expected packet loss probability

Table 3.4 State of DERs under different microgrid operating conditions.

Load RERs	Low	High
Low	• DEG: OFF • BESS: discharging or charging • FESS: discharging or charging	• DEG: ON • BESS: discharging • FESS: discharging
High	• DEG: OFF • BESS: charging • FESS: charging	• DEG: OFF • BESS: discharging or charging • FESS: discharging or charging

microgrid load is categorized into low and high, and the excess generation is used to charge ESS. In particular, these four different cases are:

- **Case 1**: The power generated by RERs is low. The load is low and consumes most of the power generated by RERs. As a result, the DEG is OFF. Stochastic RERs caused by the dynamic weather conditions lead to unbalance between supply and demand, reflected by system frequency fluctuations. These can be mitigated by the ESS via the controller whose input is the system frequency derivation, and therefore system frequency is stabilized.
- **Case 2**: The power generated by RERs is low and the load is high. It indicates that RERs cannot supply sufficient power, and therefore complementary power is provided by the DEG and the discharging ESS.
- **Case 3**: The power generated by RERs is high and the load is low. RERs can supply sufficient power, and therefore the excess power is used to charge ESS and the DEG is OFF.
- **Case 4**: The power generated by RERs is high and the load is high. The combination of RERs and ESS is sufficient to supply the total load and the DEG is disconnected.

For the above cases, the initial conditions of the DERs are listed in Table 3.4. These conditions match the realistic situations in Figure 3.4. The reference system frequency is set to 50 Hz.

3.3 HMM-Based Stability Improvement

3.3.1 On-line Smith Predictor

The Ethernet-based open communication network is mapped into five levels of congestion, i.e., very low, low, medium, high and very high, which are characterized by the DHMM as shown in Figure 3.2. Feasible scalar quantization techniques map the measured random delays into the discrete observation space of the DHMM [57–59].

The Missing Data Expectation Maximization (MDEM)-based Baum-Welch algorithm and the Viterbi algorithm are introduced to estimate the parameters of the DHMM and predict the time delay, respectively. The predictions $\hat{\tau}_k^{ca}$ and $\hat{\tau}_{k+1}^{sc}$ are used as the inputs of the on-line Smith predictor to reduce the performance loss of LFC. For comparison purpose, the Exponentially Weighted Moving Average (EWMA) method is also introduced to predict the $\hat{\tau}_k^{ca}$ and $\hat{\tau}_{k+1}^{sc}$.

3.3.1.1 Initialization of DHMM

As shown by Figure 3.2, the communication network has 5 states and the state space is denoted by $Q = \{1, 2, \cdots, 5\}$. The state in period k is q_k, $q_k \in Q$. The discrete delay observation space O has M elements and $O = \{1, 2, \cdots, M\}$. According to the uniform quantization technique, we construct M complete subintervals:

$$\left(h_L, h_U\right] = (h_0, h_1] \cup (h_1, h_2] \cup \cdots \cup (h_{M-2}, h_{M-1}] \cup (h_{M-1}, h_M] \tag{3.27}$$

where h_0 equals to h_L and h_M equals to h_U. The delay τ_k is assumed to be comprised within the interval $(h_L, h_U]$ and a new observation o_k should be defined as $o_k = l$, $o_k \in O$.

In real open communication network, time delays are not always uniformly distributed within the interval $(h_L, h_U]$. In fact, when the network has light traffic, the time delays are usually concentrated around the subintervals in (3.27) with small statistical mean. Conversely, the time delays are usually concentrated in the subintervals in (3.27) with large statistical mean in case of communication congestion.

Therefore, the distribution of time delays cannot be properly modelled by the uniform quantization technique, and a better quantization technique is needed to provide more accurate time delays predictions, and better LFC performance.

To this aim, the clustering quantization technique is more suitable for non-uniform distributed time delays. The K-means algorithm is one of the popular unsupervised clustering algorithms and partitions all observations into different clusters in which each observation belongs to the cluster with the nearest mean. This algorithm has been proved to be very effective in handling large amounts of observations.

In the initialization step, the number of clusters is pre-assigned as M. Given a set of time delay measurements $= \{\tau_1, \tau_2, \cdots, \tau_K\}$, where K is the number of measurements, the K-means clustering algorithm aims to partition the measurements into M ($N < M < K$) sets, i.e., $S = \{S_1, S_2, \cdots, S_M\}$, to minimize the within-cluster sum of square (WCSS), i.e., the sum of distance function of each point in the cluster to the center:

$$\arg\min_{S} \sum_{i=1}^{M} \sum_{\tau \in S_i} \|\tau - c_i\|^2 \tag{3.28}$$

where c_i is the centroid of elements in S_i. As the sum of squares is the squared Euclidean distance, the obtained result is intuitively the "nearest" centroid.

Thus, the quantization process based on the K-means clustering algorithm is given as follows:

- **Input**: number of clusters M and the time delay measurements $\tau = \{\tau_1, \tau_2, \cdots, \tau_K\}$
- **Step 1**: randomly generate an initial set of M centroids $c_1^{(1)}, \cdots, c_M^{(1)}$ and set maximum number of iterations (default as 100). Then, the algorithm starts.

- **Step 2:** (assignment step) assign each time delay measurement to the cluster whose means has the least WCSS in v-th iteration. Since time delay measurements and all cluster centroids are in one-dimensional Euclidean space, this chapter only uses the one-dimensional Euclidean distance:

$$S_i^{(v)} = \{\tau_k : \|\tau_k - c_i^{(v)}\| \le \|\tau_k - c_j^{(v)}\|\} \forall j, 1 \le j \le M \tag{3.29}$$

where each τ_k can only be assigned to exactly one $S_i^{(v)}$.

- **Step 3:** (update step) calculate the new means to be the centroids for the observations in the new clusters:

$$c_i^{(v+1)} = \frac{1}{|S_i^{(v)}|} \sum_{\tau_k^{ca} \in S_i^{(v)}} \tau_k \tag{3.30}$$

where the centroid is a least-squares estimator and also minimizes the WCSS objective.

- **Step 4:** Repeat Step 3 and Step 4 until the maximum number of iterations is reached.
- **Output:** All clusters $S = \{S_1, S_2, \cdots, S_M\}$ and their centroids $c_1, c_2, \cdots, c_M$.

Thus, based on the K-means clustering quantization, when $\tau_k \in S_i$, a new observation o_k is defined as $o_k = i, i \in O$ holds.

After K periods, a set of underlying network states $\boldsymbol{q} = \{q_1, q_2, \cdots, q_K\}$ and a set of time delays $\boldsymbol{\tau} = \{\tau_1, \tau_2, \cdots, \tau_K\}$ is obtained. Then, a set of corresponding observations $\boldsymbol{o} = \{o_1, o_2, \cdots, o_K\}$ is created by applying the scalar quantization techniques to $\boldsymbol{\tau}$. The sets $\boldsymbol{q}$, $\boldsymbol{\tau}$ and $\boldsymbol{o}$ are the fundamental elements of the DHMM for the channels of the open communication network [57].

During each signal transmission in the channel, the network transitions among states or stays in the current state [53, 54]. These state transitions define a Markov chain across the congestion states, as shown in Figure 3.7. The Markov chain takes values from $Q = \{1, 2, \cdots, N\}$ and the state transition probability matrix is $P = [p_{ij}]$. The transition probability from the state i in period k to the state j in period $k + 1$ is

$$p_{ij} = \Pr\{q_{k+1} = j | q_k = i\} \tag{3.31}$$

where $p_{ij} \ge 0$, $i, j \in Q$, and $\sum_{j \in Q} p_{ij} = 1$. The first period ($k = 1$) is characterized by the initial state distribution $\boldsymbol{\pi} = [\pi_1, \cdots, \pi_i, \cdots, \pi_N]$, where $\pi_i = \Pr\{q_1 = i\}$ and $\sum_{i \in Q} \pi_i = 1$.

Figure 3.7 Schematic of DHMM: Observed time delays are generated by the underlying and unobserved Markov chain representing the network states.

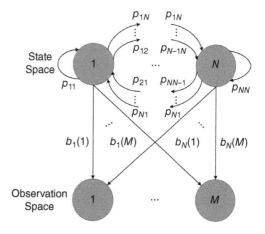

As the time delays are determined by network states, the probability of a particular observation at period k, e.g., $o_k = l$, given that the network is in state i, i.e., $q_k = i$, is described by Figure 3.7:

$$b_i(l) = \Pr\{o_k = l | q_k = i\} \tag{3.32}$$

(3.32) satisfies the stochastic constraints $b_i(l) \geq 0$ and $\sum_{l \in O} b_i(l) = 1$. Thus, the complete set of parameters for all observation distributions can be denoted by $B = [b_i(l)]$, where $i \in Q$ and $l \in O$.

Based on the above definitions, the DHMM is expressed as:

$$\lambda = (N, M, \pi, P, B) \tag{3.33}$$

Traditionally, N and M are known in advance via experience or historical data [53, 54]. Thus, (3.33) simplifies as

$$\lambda = (\pi, P, B) \tag{3.34}$$

The probabilistic relationship between the network states and networked-induced delays in both channels can be described by the DHMM in (3.34) and is critical to the delay prediction.

3.3.1.2 Parameter Estimation of DHMM

The parameters (π, P, B) of DHMM for the network state and the time delays in (16) need to be estimated using the set of time delay observations $o = \{o_1, o_2, \cdots, o_K\}$ obtained in the K periods. To this aim, the Expectation Minimization (EM) is widely used for its efficiency in finding the maximum likelihood estimation (MLE) of the parameters of a statistical model given observations [51]. Therefore, the MLE of the parameters (π, P, B) based on the EM algorithm is:

$$\lambda^* = (\pi^*, P^*, B^*) = \arg\max_{\lambda} \Pr(o | \lambda) \tag{3.35}$$

where $\Pr(o | \lambda)$ is the likelihood function which needs to be maximized during the EM algorithm.

The EM algorithm aims to find optimal λ^* for the DHMM which can most likely generate the set of the time delay observations $o = \{o_1, o_2, \cdots, o_K\}$. $\Pr(o | \lambda)$ in (3.35) is actually an incomplete data likelihood function, i.e., does not employ the underlying data (network state q) in the optimization process, and can be rewritten as $\lg \Pr(o | \lambda)$ due to the monotonicity of logarithmic function.

To solve the missing data problem, based on the MDEM-based Baum-Welch algorithm, the hidden network state q is considered, and a complete data log-likelihood function $\lg \Pr(o, q | \lambda)$ is achieved. Thus, the cost function used in the Baum-Welch algorithm is given as

$$G(\lambda, \lambda') = E\{\lg \Pr(o, q | \lambda) | o, \lambda'\} \tag{3.36}$$

where λ' is the optimal parameter estimation obtained in previous iteration and used to evaluate the expectation. λ is the new parameter estimation to be optimized for the purpose of increasing G.

As the underlying network state q is a discrete sequence which takes values from Q, G can be modified as

$$G(\lambda, \lambda') = \sum_{q \in Q} \lg \Pr(o, q | \lambda) \Pr(q | o, \lambda')$$

$$= \sum_{q \in Q} \lg \Pr(o, q | \lambda) \Pr(o, q | \lambda') / \Pr(o | \lambda') \tag{3.37}$$

$$= \frac{1}{\Pr(o | \lambda')} \sum_{q \in Q} \lg \Pr(o, q | \lambda) \Pr(o, q | \lambda')$$

As the cost function in (3.37) is maximized only with respect to λ, i.e., $\Pr(o | \lambda')$ can be neglected, the likelihood function can be rewritten as

$$G(\lambda, \lambda') = \sum_{q \in Q} \lg \Pr(o, q | \lambda) \Pr(o, q | \lambda') \tag{3.38}$$

Theorem 3.1 If $G(\lambda, \lambda') \geq G(\lambda', \lambda')$, $\Pr(o | \lambda) \geq \Pr(o | \lambda')$ holds.
Proof: Based on [57, 66], the following relationship holds

$$\log \Pr\{o | \lambda\} - \log \Pr\{o | \lambda'\} = \frac{1}{\Pr\{o | \lambda'\}} \{G(\lambda, \lambda') - G(\lambda', \lambda')\} \tag{3.39}$$

Thus, when $G(\lambda, \lambda') \geq G(\lambda', \lambda')$, there exists $\log \Pr\{o | \lambda\} - \log \Pr\{o | \lambda'\} \geq 0$, i.e., $\Pr\{o | \lambda\} \geq \Pr\{o | \lambda'\}$.

Theorem 3.1 illustrates the reason that the MDEM-based Baum-Welch algorithm can maximize the likelihood function of the DHMM for time delays. Each iteration in the MDEM-based Baum-Welch algorithm consists of two steps, which are given as follows

- **E-step**: compute $G(\lambda, \lambda') = \sum_{q \in Q} \lg \Pr(o, q | \lambda) \Pr(o, q | \lambda')$.

$$G(\lambda, \lambda') = \sum_{q \in Q} \lg \pi_{q_1} \Pr(o, q | \lambda') + \sum_{q \in Q} \left(\prod_{k=2}^{K} \lg p_{q_{k-1} k} \right) \Pr(o, q | \lambda')$$

$$\sum_{q \in Q} \left(\prod_{k=1}^{K} \lg b_{q_k}(o_k) \right) \Pr(o, q | \lambda') \tag{3.40}$$

- **M-step**: find a local maximum λ^* that maximizes $G(\lambda, \lambda')$, i.e., $\lambda^* = \arg\max_\lambda G(\lambda, \lambda')$.

As shown by (3.40), the optimization problem of maximizing $G(\lambda, \lambda')$ are independently split into three problems of deriving maximum values for three terms: π^*, P^* and B^*. These values are achieved by solving its derivative via three procedures, i.e., forward procedure, backward procedure and update procedure, given by the MDEM-based Baum-Welch algorithm [66]. The results of π^*, P^* and B^* for the DHMM have already been given by [57].

These two steps, i.e., E-step and M-step, are repeated until a desired level of convergence, i.e., $|\lambda^* - \lambda'| < \varepsilon$, is reached. The final result would converge to an extremum of the likelihood function.

However, as λ consists of π, P and B, it is very hard to compute the distance between λ^* and λ'. Notice that $\lambda^* = \arg\max_\lambda \Pr(o | \lambda)$ and $\Pr(o | \lambda) = \sum_{q \in Q} \Pr(q, o | \lambda)$, whose main component is $\max_q \Pr(q, o | \lambda)$, thus it is more practical to decide whether the distance between $\max_q \Pr(q, o | \lambda^*)$ and $\max_q \Pr(q, o | \lambda')$ is smaller than ε or not. The most convenient method for computing the $\max_q \Pr(q, o | \lambda)$ is the Viterbi algorithm [67].

Theorem 3.2   If $|\max_q \Pr(q, o | \lambda^*) - \max_q \Pr(q, o | \lambda')| < \varepsilon$, $|\lambda^* - \lambda'| < \varepsilon$ holds.
Proof: The proof can be referred to [57, 67].

Therefore, when $|\lambda^* - \lambda'| < \varepsilon$ is reached, the MDEM-based Baum-Welch algorithm halts. The most likely hidden state sequence is computed and used in the time delay prediction.

3.3.1.3 Delay Prediction via DHMM
Denoting the prediction of τ_k as $\hat{\tau}_k$, the prediction of q_k as $\hat{q}_k$ and the prediction of o_k as $\hat{o}_k$, where $\hat{q}_k \in Q$ and $\hat{o}_k \in O$, the method for quantifying $\hat{\tau}_K$ is based on the uniform and on the clustering quantization technique used in the estimation of the DHMM.

The technique of quantifying $\hat{\tau}_{K+1}$ via the uniform quantization is detailed as bellow:

- Use λ^* and q_K^* to predict the most likely network state $\hat{q}_{K+1}$:

$$\hat{q}_{K+1} = \max_{1 \leq j \leq N}(q_{K,j}^*) \tag{3.41}$$

- Use λ^* and $\hat{q}_{K+1}$ to predict the most likely $\hat{o}_{K+1}$:

$$\hat{o}_{K+1} = \max_{1 \leq l \leq M}(b_{\hat{q}_{K+1}}^*(l)) \tag{3.42}$$

- Based on the uniform quantization technique and $\hat{o}_{K+1}$:

$$\hat{\tau}_{K+1} = (h_{\hat{o}_{K+1}} - h_{\hat{o}_K})/2 \tag{3.43}$$

i.e., $\hat{\tau}_{K+1}$ equals to the midpoint of the subinterval which comprises $\hat{o}_{K+1}$.

The technique of quantifying $\hat{\tau}_{K+1}$ via the K-means clustering quantization is detailed as bellow:

- The first two steps are same as the steps presented in (3.41) and (3.42).
- Based on the K-means clustering quantization technique and $\hat{o}_{K+1}$:

$$\hat{\tau}_{K+1} = c_{\hat{o}_{K+1}} \tag{3.44}$$

i.e., $\hat{\tau}_{K+1}$ equals to the centroid of the cluster which comprises $\hat{o}_{K+1}$.

In the period K, the control center has S-C time delay measurements $\tau_1^{sc}, \tau_2^{sc}, \cdots, \tau_K^{sc}$ and C-A time delay measurements $\tau_1^{ca}, \tau_2^{ca}, \cdots, \tau_{K-1}^{ca}$. Thus, according to the scale quantization technique, corresponding observation $o_1^{sc}, o_2^{sc}, \cdots, o_K^{sc}$ and $o_1^{ca}, o_2^{ca}, \cdots, o_{K-1}^{ca}$ can be obtained. Thus, using the MDEM-based Baum-Welch algorithm, optimal parameters λ^* for the DHMM can be derived.

Finally, by using the Viterbi algorithm, the most likely network state sequence $q_1^{sc^*}$, $q_2^{sc^*}, \cdots, q_K^{sc^*}$ and $q_1^{ca^*}, q_2^{ca^*}, \cdots, q_{K-1}^{ca*}$ can be generated. Then, the prediction $\hat{\tau}_K^{ca}$ and $\hat{\tau}_{K+1}^{sc}$ can be obtained.

In the EWMA method, $\hat{\tau}_{K+1}$ is predicted as [49]:

$$\hat{\tau}_{K+1} = \alpha[\tau_K + (1-\alpha)\tau_{K-1} + \cdots + (1-\alpha)^n \tau_{K-n-1}] \tag{3.45}$$

where α is the weight and n is the window length.

3.3.1.4 Smith Predictor Structure
The delayed control process can be effectively handled by the Smith predictor if the information on time delays is known [41, 69]. In this chapter, the time delays of the network

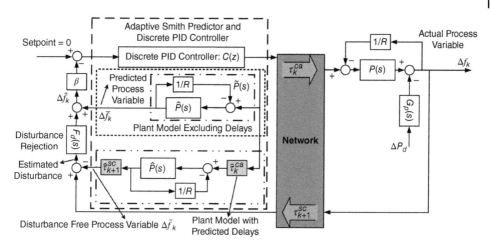

Figure 3.8 The illustrative WAPS with the DHMM-based Smith predictor.

τ_k are measured by the timestamp technique of Section 2.3, and $\hat{\tau}_{K+1}$ is predicted by the DHMM or the EWMA method in this section, and implemented into the Smith Predictor shown by the dashed box in Figure 3.8.

As shown in Figure 3.8, $P(s)$ denotes the transfer function of the real plant in Figure 3.1 and equals to $G_G(s)G_T(s)G_P(s)$. $\hat{P}(s)$ is the nominal model of the WAPS and $\hat{P}(s) = P(s)$. Taking into account the impact of droop characteristic on $\hat{P}(s)$, the internal model $\tilde{P}(s)$ is

$$\tilde{P}(s) = \frac{G_G(s)G_T(s)G_P(s)}{1 + G_G(s)G_T(s)G_P(s)/R}. \tag{3.46}$$

The Smith predictor uses an internal model, $\tilde{P}(s)$, to predict the delay-free process variable, $\Delta\hat{f}_k$, i.e., the predicted process variable in Figure 3.8. To prevent drifting and reject external disturbance, the Smith predictor subtracts the disturbance-free process variable $\Delta\bar{f}_k$, which takes predicted time delays, i.e., τ_k^{ca} and τ_{k+1}^{sc}, into account, from the actual process variable, yielding an estimate of the disturbances.

The estimated disturbance is fed back through a filter $F_d(s)$ to improve the disturbance rejection performance. Based on [18], the $F_d(s)$ is as follows:

$$F_d(s) = \frac{a_n s^n + \cdots + a_1 s + 1}{(\gamma s + 1)^n} \tag{3.47}$$

where n is the number of the poles of $\tilde{P}(s)$ such that $F_d(s)$ needs to cancel, and γ is a tuning parameter for disturbance rejection.

Defining the poles to be canceled as $r_1, \cdots, r_n$, the coefficients $a_1, \cdots, a_n$ satisfy:

$$(1 - \tilde{P}(s)F_d(s))\Big|_{s=r_1,\cdots,r_n} = 0. \tag{3.48}$$

Finally, the discrete PID controller is driven by a delay-free signal $\Delta\hat{f}_k$, i.e., the superposition of the predicted process variable $\Delta\hat{f}_k$ and the estimated disturbance $\Delta\tilde{f}_k$ processed by the filter $F_d(s)$. A small error signal, representing the error caused by time delays and disturbance, is added to finely adjust the control signal to remove the closed loop difference between the reference model and the actual plant, and thus stabilizes the WAPS in the face of communication delays.

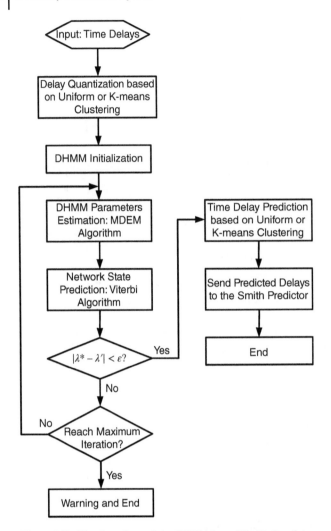

Figure 3.9 The flowchart of the DHMM-based Smith Predictor.

The application of the proposed DHMM-based Smith predictor to the WAPS with communication delays is represented in Figure 3.9.

3.3.2 Delay Predictions

To demonstrate to the accuracy of the DHMM method, a simulation-based experiment is performed on the TrueTime 2.0 [51, 54, 59].

3.3.2.1 Settings of DHMM

The interval $(h_L, h_U]$ of time delays is divided into seven complete subintervals based on the dataset of time delays [35], i.e., $M = 7$ and $O = \{1, 2, 3, 4, 5, 6, 7\}$ [59]. Given the parameters for the Ethernet in Table 3.1, the C-A time delays are comprised within the interval

(0.048,0.158] and the S-C time delays are within the interval (0.037,0.126]. For the EWMA, $\alpha = 0.8$ and $n = 10$ [52].

Applying the uniform quantization technique to the C-A time delays, (3.27) is determined as follows:

$$(0.048, 0.160] = (0.048, 0.064] \cup (0.064, 0.080] \cup (0.080, 0.096] \cup (0.096, 0.112]$$

$$\cup (0.112, 0.128] \cup (0.128, 0.144] \cup (0.144, 0.160]. \tag{3.49}$$

Based on the MDEM-based Baum-Welch algorithm, the DHMM parameters estimated from one sample containing 100 C-A time delay measurements, i.e., the length of most recent time delays is 100, are

$$\pi_1^* = (0\,0\,0\,0\,1) \tag{3.50}$$

$$P_1^* = \begin{pmatrix} 0.85 & 0.15 & 0 & 0 & 1 \\ 0 & 0.96 & 0 & 0.04 & 0 \\ 0 & 0 & 1 & 0 & 0 \\ 0 & 0.05 & 0 & 0.89 & 0.06 \\ 0 & 0 & 0.07 & 0 & 0.93 \end{pmatrix} \tag{3.51}$$

$$B_1^* = \begin{pmatrix} 0 & 1 & 0 & 0 & 0 & 0 & 0 \\ 0 & 0 & 1 & 0 & 0 & 0 & 0 \\ 0 & 0 & 0.07 & 0.49 & 0.44 & 0 & 0 \\ 0 & 0 & 0 & 0 & 1 & 0 & 0 \\ 0 & 0 & 0 & 0.23 & 0.77 & 0 & 0 \end{pmatrix} \tag{3.52}$$

where $\lambda_1^* = (\pi_1^*, P_1^*, B_1^*)$ are the estimation results derived from the uniform quantization technique.

In order to make comparisons, the estimation results $\lambda_2^* = (\pi_2^*, P_2^*, B_2^*)$ derived from the K-means clustering quantization technique are:

$$\pi_2^* = (0\,0\,0\,0\,1) \tag{3.53}$$

$$P_2^* = \begin{pmatrix} 0.83 & 0.17 & 0 & 0 & 0 \\ 0 & 0.15 & 0.80 & 0 & 0.05 \\ 0 & 0.55 & 0.41 & 0 & 0.04 \\ 0 & 0 & 0 & 1 & 0 \\ 0 & 0 & 0.02 & 0.03 & 0.95 \end{pmatrix} \tag{3.54}$$

$$B_2^* = \begin{pmatrix} 0.80 & 0.20 & 0 & 0 & 0 & 0 & 0 \\ 0 & 1 & 0 & 0 & 0 & 0 & 0 \\ 0 & 0 & 0.52 & 0.48 & 0 & 0 & 0 \\ 0 & 0 & 0 & 0.23 & 0.46 & 0.31 & 0 \\ 0 & 0 & 0 & 0 & 0 & 0.52 & 0.48 \end{pmatrix} \tag{3.55}$$

where the seven centroids are: 0.072, 0.082, 0.086, 0.092, 0.102, 0.112 and 0.120.

3.3.2.2 Prediction Comparison

The results of the estimation processes show that λ_1^* and λ_2^* are different, because different quantization techniques produce different observations $\{o_1, \cdots, o_{100}\}$.

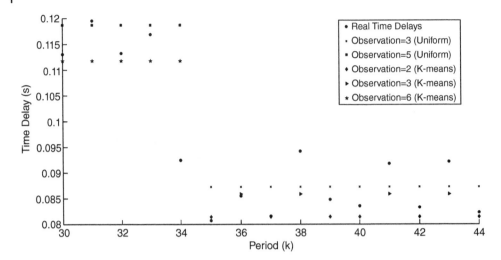

Figure 3.10 The C-A time delay observations derived from different quantization techniques.

Figure 3.10 displays the observations for the periods 30 to 44. In the period 40, the delay measurement, i.e., 0.084, is comprised within the third subinterval (0.080,0.096] based on the uniform quantization technique, and, therefore, $o_{40} = 3$. Conversely, the same delay measurement, i.e., 0.084, is closer to the second cluster centroid, i.e., 0.082, for the K-means quantization technique, and, therefore, $o_{40} = 2$.

Therefore, different quantization techniques may produce different observation values for the same delay measurement. Since the centroids in the K-means quantization technique are statistically inferred from the distribution of time delays, the observations derived from the K-means quantization technique can better track the actual time delays as compared to the uniform quantization technique as shown by Figure 3.10.

The tracking ability of the K-means quantization technique results in more accurate predictions as compared to the uniform quantization technique. Based on the estimated λ_1^*, λ_2^* and the proposed time delay prediction method defined by (3.41) - (3.44), the corresponding prediction results of the EWMA and DHMM method are shown in Figure 3.11.

The mean square error (MSE) of the EWMA, Uniform-based DHMM and K-means-based DHMM are 0.0636, 0.0563 and 0.0478, respectively. The prediction of EWMA lags behind the real time delay, because it is based on the moving average of actual time delays. The prediction of DHMM relies on the most likely communication network state and stays constant until the network state changes. Figure 3.11 shows that the K-means-based DHMM provides more accurate predictions than the uniform-based DHMM because the observations of the K-mean quantization technique are statistically inferred.

Figure 3.12 displays the most likely network states estimated by the Viterbi algorithm. For the uniform quantization technique, the most likely network state in period 40 is $q_{40} = 2$. Based on (3.41) and P_1^*, the predicted network state in period 41 is $\hat{q}_{41} = 2$. Based on (3.42) and B_1^*, the predicted observation in period 41 is $\hat{o}_{41} = 3$. Therefore, the predicted time delay in the period 41 determined by (3.43) is 0.0873s.

For the K-means quantization technique, the most likely network state of period 40 is $q_{40} = 2$, and, thereby, the predicted network state of period 41 is $\hat{o}_{41} = 3$, based on (3.41)

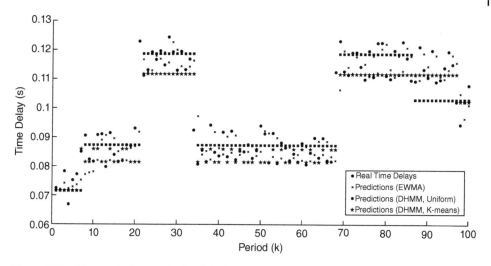

Figure 3.11 The prediction results for C-A time delays (Source: Mo and Sansavini, 2021 [309]).

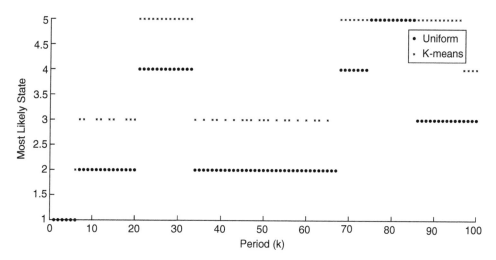

Figure 3.12 Most likely network states derived from the two different quantization techniques.

and P_2^*. Based on (3.42) and B_2^*, the predicted observation of period 41 is $\hat{o}_{41} = 3$, and, therefore, the predicted time delay in period 43 belongs to the third centroid, i.e., 0.0859s, based on (3.44).

Table 3.5 shows the statistical properties of the MSE of the EWMA and DHMM method for C-A and S-C delays, computed from 1000 samples each containing 100 delay measurements to ensure accuracy.

The DHMM outperforms the EWMA in the delay prediction because of the smaller mean MSE. Compared with the EWMA, the uniform-based DHMM reduces the MSE of C-A and S-C delay predictions by 3.47% and 3.72%. Compared with the EWMA, the K-means-based DHMM reduces the MSE of C-A and S-C delay predictions by 7.19% and 7.26%.

Table 3.5 Statistical properties of MSE of delay predictions.

Delay	Method	μ	95% Confidence Interval
C-A	EWMA	0.0640	[0.0630, 0.0649]
C-A	Uniform	0.0605	[0.0596, 0.0614]
C-A	K-means	0.0594	[0.0585, 0.0603]
S-C	EWMA	0.0647	[0.0637, 0.0658]
S-C	Uniform	0.0610	[0.0597, 0.0623]
S-C	K-means	0.0606	[0.0597, 0.0617]

Delay	Method	σ	95% Confidence Interval
C-A	EWMA	0.0151	[0.0145, 0.0158]
C-A	Uniform	0.0145	[0.0139, 0.0152]
C-A	K-means	0.0143	[0.0137, 0.0150]
S-C	EWMA	0.0169	[0.0162, 0.0176]
S-C	Uniform	0.0158	[0.0149, 0.0167]
S-C	K-means	0.0147	[0.0140, 0.0154]

Indeed, the prediction of EWMA depends on recent delays while the prediction of DHMM depends on the current network state, which is the underlying process causing delays. Because the observations of K-means-based DHMM is able to capture the statistical distribution of time delays, the MSE of the K-means-based DHMM is smaller than the one of the uniform-based DHMM.

3.3.3 Performance of Smith Predictor

3.3.3.1 Settings of Smith Predictor
At $t = 35s$, a positive load disturbance of 0.1 pu is added to the WAPS with communication delays described in Table 3.1 [17, 24]. We study two different cases, i.e., the LFC via discrete PI controller derived from the delay margin method [17] and LFC via robust PID controller derived using genetic algorithms (GA) [35]. These two controllers are used by most real power systems [17–24, 35, 46].

The internal plant model $\widetilde{P}(s)$ with droop characteristic is

$$\widetilde{P}(s) = \frac{100}{30s^3 + 403s^2 + 1040s + 2100} \tag{3.56}$$

where $\widetilde{P}(s)$ has a pair of complex poles $-1.302 \pm 2.184i$.

As the response is oscillatory, the filter $F_d(s)$ given in (3.47) is used to cancel the effect of the oscillatory poles and reject the disturbance. Based on (3.48), with $n = 2$ and the designed $\gamma = 1.2$, the a_2 and a_1 in $F_d(s)$ are derived as 0.154 and 0.403. Therefore, the $F_d(s)$ is expressed as

$$F_d(s) = \frac{0.154s^2 + 0.403s + 1}{(1.2s + 1)^2} \tag{3.57}$$

To quantitatively evaluate the influence of communication delays on the LFC of the single-area WAPS, this chapter employs an aggregated indicator, which measures frequency fluctuations and control efforts:

$$J = \int_0^T [\omega \cdot (\Delta f)^2 + (\Delta u)^2])dt \tag{3.58}$$

where $\omega = 2 \times 10^3$ is a normalized factor for the deviation of system frequency.

3.3.3.2 Analysis of Case 1

For the time-varying delays, a discrete PI controller ($K_p = 1$ and $T_i = 0.6$) can tolerate a maximal allowable delay of 0.281s, based on Table III in [17]. As the upper bound of the cumulative time-varying delays in the C-A channel and the S-C channel is 0.280s, this PI controller can preserve the stability of the LFC.

Figure 3.13 shows the LFC performance of the WAPS under the delay-margin-based PI controller with and without the proposed Smith predictor and the designed disturbance filter. The PI controller takes more than 200s to stabilize the system frequency and the designed disturbance filter cannot reduce frequency oscillations as illustrated by Figure 3.13 (a). The Smith predictor can largely reduce the frequency deviations and settling time as shown by Figure 3.13 (b). Figure 3.13 (c) shows that the use of the Smith predictor and the designed disturbance filter can provide enough disturbance rejection, and, therefore, further reduce the active frequency oscillations, which could damage generators, trip tie lines, and increase wear and tear on power system components.

Figure 3.13 (b) and Figure 3.13 (c) show that the Smith predictor based on the DHMM reduces the frequency drop and the setting time as compared to the EWMA. Additionally, the accuracy of the delay prediction by K-means-based DHMM (Table 3.5) enhances the system LFC performance against the positive load disturbance as compared to the uniform-based DHMM.

Table 3.6 shows the statistical properties of the indicator (3.58) for different scenarios under case 1, computed using 1000 samples. The Smith predictor greatly reduces frequency fluctuations as compared to the original WAPS.

The comparison of Scenario 3 and 4, and of Scenario 3 and 5, shows that the uniform-based DHMM and K-means-based DHMM enhances the performance of the LFC by 4.6% and by 6.9%, respectively, as compared to the EWMA, according to the indicator in (3.58). The comparison of Scenario 3 and 6, of Scenario 4 and 7, and of Scenario 5 and 8, shows that use of the disturbance filter further improves the performance of the LFC by 33.3%, by 36.8% and by 36.5%, respectively, as compared to scenarios without using the filter.

3.3.3.3 Analysis of Case 2

For the robust PID controller, K_p, T_i and T_d are set to 0.20, 0.20 and 1.01, via the GA-based robust LFC strategy [35, 307].

Figure 3.14 shows the LFC performance of the WAPS under the robust PID controller with and without the proposed Smith predictor and the designed disturbance filter. The comparison of Figure 3.13 (a) and Figure 3.14 (a) shows that the robust PID controller is better than the PI controller in stabilizing system frequency. Consistently, the introduction of Smith predictor is effective in eliminating the system frequency oscillation and reducing the settling time. The usage of disturbance filter can further improve the LFC performance.

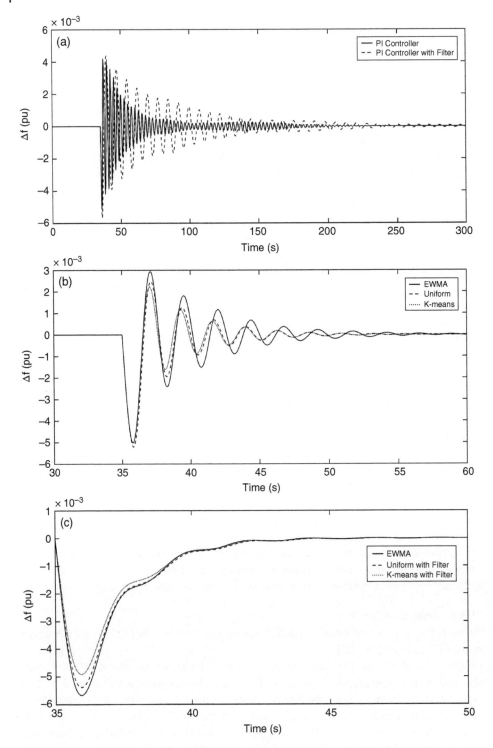

Figure 3.13 LFC performance of WAPS (a) without Smith predictor, (b) with Smith predictor and (c) with Smith predictor and disturbance filter.

Table 3.6 Statistical properties of aggregated indicator of the case 1.

Scenario	Method	μ	95% Confidence Interval
1	PI controller	4821	[4745, 4897]
2	PI controller with Filter	13284	[12768, 13800]
3	EWMA without Filter	2698	[2696, 2701]
4	Uniform-based DHMM without Filter	2573	[2570, 2575]
5	K-means-based DHMM without Filter	2512	[2515, 2521]
6	EWMA with Filter	1800	[1784, 1816]
7	Uniform-based DHMM with Filter	1627	[1622, 1631]
8	K-means-based DHMM with Filter	1595	[1590, 1599]

Scenario	Method	σ	95% Confidence Interval
1	PI controller	1225	[1173, 1281]
2	PI controller with Filter	1835	[1535, 2281]
3	EWMA without Filter	39.44	[37.79, 41.25]
4	Uniform-based DHMM without Filter	33.78	[34.28, 37.42]
5	K-means-based DHMM without Filter	34.12	[32.69, 33.67]
6	EWMA with Filter	23.54	[24.47, 26.71]
7	Uniform-based DHMM with Filter	23.46	[22.47, 24.53]
8	K-means-based DHMM with Filter	22.35	[21.41, 23.37]

Even though the robust LFC strategy is well designed by [35], the Smith predictor can still improve LFC performance, based on the comparison of Figure 3.14 (a) and Figure 3.14 (b). The use of disturbance filter also works for the robust PID controller and further reduces frequency oscillations, based on the comparison of Figure 3.14 (b) and Figure 3.14 (c).

Table 3.7 shows the statistical properties of the indicator (3.58) for different scenarios and for Case 2, computed using 1000 samples. The comparison of Scenario 3, 4 and 5, and of Scenario 6, 7 and 8 presented in Table 3.7, shows that the K-means-based DHMM has the smallest indicator value and the best performance in damping the WAPS frequency oscillations, because it can provide the best time delay predictions as compared to the EWMA and the uniform-based DHMM.

The comparison between Scenario 1 in Table 3.6 and Scenario 1 in Table 3.7 shows that the robust PID is more effective in reducing frequency fluctuations than the delay-margin based PI controller and improves the value of aggregated indicator by 14.1%. Indeed, the robust PID controller carries out an additional step to search for its optimal parameters, to achieve compromise between the LFC performance and the delay margin [35], besides just computing the delay margin.

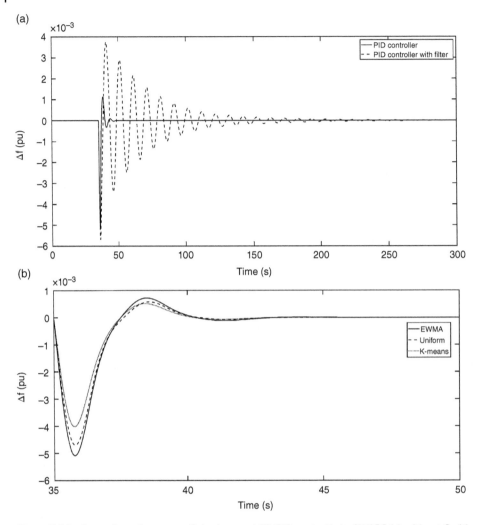

Figure 3.14 Dynamic performance of single-area LFC (PID controller) of WAPS (a) without Smith predictor, (b) with Smith predictor and (c) with Smith predictor and disturbance filter.

As a result, the line-by-line cross comparison between Table 3.6 and Table 3.7 shows that the robust PID controller improves the value of aggregated indicator of real systems by 6.8% to 14.5%, as compared to the delay-margin based PI controller. Since the delay-margin based PI controller and the robust PID controller are widely implemented in practice [17, 24, 35], it is significant to compare the results of our method against the real case studies, at which above two controllers are used.

The comparison of Scenario 1 and 8 in Table 3.6, and of Scenario 1 and 8 in Table 3.7, shows that our method, i.e., the Smith predictor with disturbance filter, reduces the value of aggregated indicator of real systems by 66.9% and by 64.1%, respectively, as compared to the real case studies.

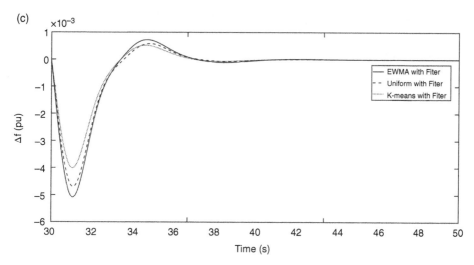

Figure 3.14 (*Continued*)

Table 3.7 Statistical properties of aggregated indicator of the case 2.

Scenario	Method	μ	95% Confidence Interval
1	PI controller	4144	[4136, 4153]
2	PI controller with Filter	11403	[10994, 11813]
3	EWMA without Filter	2307	[2304, 2309]
4	Uniform-based DHMM without Filter	2219	[2217, 2221]
5	K-means-based DHMM without Filter	2154	[2153, 2157]
6	EWMA with Filter	1549	[1545, 1552]
7	Uniform-based DHMM with Filter	1516	[1513, 1519]
8	K-means-based DHMM with Filter	1487	[1484, 1490]

Scenario	Method	σ	95% Confidence Interval
1	PI controller	136.0	[130.3, 142.3]
2	PI controller with Filter	1456	[1218, 1810]
3	EWMA without Filter	29.27	[27.76, 30.94]
4	Uniform-based DHMM without Filter	28.30	[27.11, 29.59]
5	K-means-based DHMM without Filter	27.38	[26.23, 28.63]
6	EWMA with Filter	24.48	[23.45, 23.60]
7	Uniform-based DHMM with Filter	24.10	[23.10, 23.21]
8	K-means-based DHMM with Filter	23.78	[22.78, 24.87]

3.4 Stability Enhancement of Illustrative WAPS

3.4.1 Eigenvalue Analysis and Delay Impact

According to the state-space model of the microgrid given in Section 3.1.2.3, the eigenvalue analysis is conducted following the procedure detailed in [86, 106, 107]. Figure 3.15 compares the effect of the time delay in the proposed MOPSO-based PID controller with the case of a PI controller [304], when packet dropout is neglected, and the period is 0.01 s.

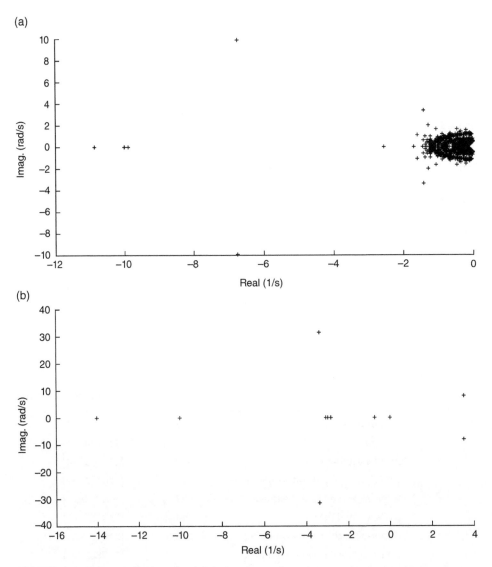

Figure 3.15 Eigenvalue spectrum of the microgrid in the presence of time delays with (a) the proposed MOPSO-based PID controller when the time delay increases from 0 to 390 ms, and (b) the PI controller when the time delay increases from 0 to 24 ms.

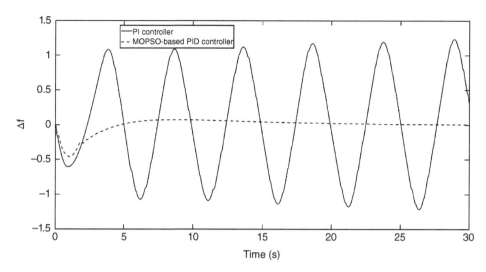

Figure 3.16 System frequency responses of (a) the PI controller with 20 ms time delay, and (b) the MOPSO-based PID controller with 390 ms time delay.

Figure 3.15 (a) shows the MOPSO-based PID controller eigenvalue spectrum when the time delay is as large as 390 ms (maximum allowable time delay). As demonstrated, the system with MOPSO-PID controller remains stable in the presence of large time delays. The advantage of MOPSO-PID controller over the PI controller [304] is better represented by Figure 3.15 (b); for the latter, the instability can occur when the time delay is as small as 20 ms.

As indicated by [106, 107], typical time delays in reality can be in the order of 100-300 ms. Therefore, the MOPSO-PID controller is effective and robust in the reliable management of RERs in the microgrid to ensure stable system frequency.

Figure 3.16 shows the system frequency curves of the PI controller [304] and the proposed MOPSO-based PID controller in the presence of 20 ms and 390 ms, respectively. Consistently with the eigenvalue analysis shown in Figure 3.15 (b), the time delay causes instability in the PI controller. By using the MOPSO-based PID controller, the system stability greatly improves against delays of the feedback frequency signal and of the control signal.

3.4.2 Sensitivity Analysis of Network Parameters

For illustrating purpose, the maximum permissible instantaneous frequency deviation is set to ±0.8 Hz according to [308]. The parameters of the discrete PID controller are tuned as $K_P = 0.52$, $K_I = 2.21$ and $K_D = 1.00$ [74, 79]. Five combinations of network configurations, i.e., BWShare, T_i and L_d, are taken into account to investigate multiple communication scenarios in the Ethernet and in the hybrid network.

For comparison purpose, the reliability and objective values of the integrated system with perfect communication, i.e., no time delays and no packet dropouts, are $R = 94.48\%$ and $E[J(\vec{x})] = [373.37, 2.82 \times 10^4]$.

Figure 3.17 presents the impact of the length of data traffic of the interference node on the reliability of the integrated system with the Ethernet ([0.2, 0.005, 5%]). The length of data

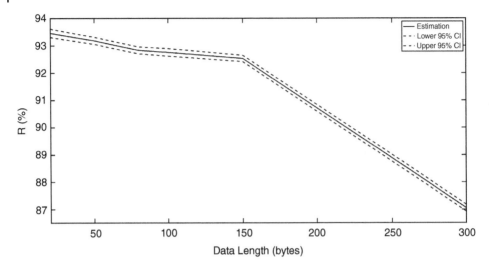

Figure 3.17 System reliability as a function of the length of data traffic.

traffic is from 20 to 200 bytes. As expected, the system reliability decreases as the length increases. It is in line with our common knowledge that a larger data packet consumes more bandwidth of the communication channel, which deteriorates network conditions and results in larger time delays and more packet dropouts.

Figure 3.18 illustrates the relationship between the reliability of the integrated system with Ethernet, and the communication configurations $[L_d, \text{BWShare}, T_i]$. Figure 3.18 (a) shows that the AGC becomes unstable and the system reliability degrades significantly if $L_d \geq 40\%$, the controller does not have sufficient observations of the system frequency measurement and is unable to compute correct control signals. The case (BWShare=0.1 & $T_i = 0.01$) in Figure 3.18 (a), has the largest reliability due to light congestion. In this case, data collisions between the PMU and the interference node are fewer than for other cases with smaller T_i. In the case (BWShare=0.2 & $T_i = 0.005$), the interference node sends out disturbing traffic every 0.005 s and consumes 20% of the network bandwidth, therefore data collisions between the PMU and the interference node increase. As a result, increased time delays degrade system reliability as compared to the other two cases of Figure 3.18 (a).

Increasing the activity level of the interference user and the used bandwidth leads to increasing data collisions, and therefore network congestion, which causes longer time delays. Thus, system reliability decreases, i.e., comparing case (BWShare=0.1 & $L_d = 0.1$) and case (BWShare=0.2 & $L_d = 0.1$) in Figure 3.18 (c), and comparing case ($T_i = 0.01$ & $L_d = 0.1$) and case ($T_i = 0.005$ & $L_d = 0.1$) in the Figure 3.18 (b), respectively. On the other hand, if $T_i \geq 0.01$, the competition for sending data packet between the interference node and the PMU reduces, network conditions improve and thus the system reliability increases in Figure 3.18 (b) and Figure 3.18 (c).

Table 3.8 provides the estimated reliability of the integrated system with Ethernet and with hybrid network. Such values of communication parameters are selected because they can generate 10 ms to 2 s time delays, and losses of 1% to 10% of the overall packet stream,

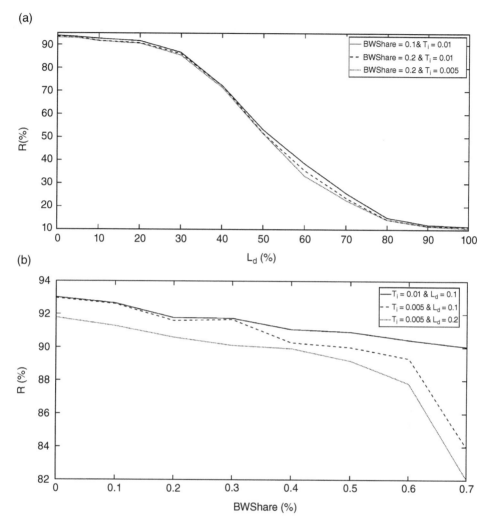

Figure 3.18 System reliability as a function of the variable of the network configuration (a) L_d, (b) BWShare and (c) T_i.

which can occur at realistic open communication networks [43, 283]. A negative correlation between system reliability and $E[J(\vec{x})]$ can be inferred from Table 3.8.

This is expected because when the system reliability is high, most of the system frequency measurements are within the tolerance interval, which thereby leads to a small value of the objective function $E[J_1(\vec{x})]$. Table 3.8 also shows that the Ethernet architecture ensures higher reliable to the integrated system as compared to the hybrid architecture. This is expected because the data exchange between the AP and the Remote Terminal Unit (RTU) in the hybrid architecture is provided by 802.11b/g, which introduces additional time delays and packet dropouts, and results in lower system reliability.

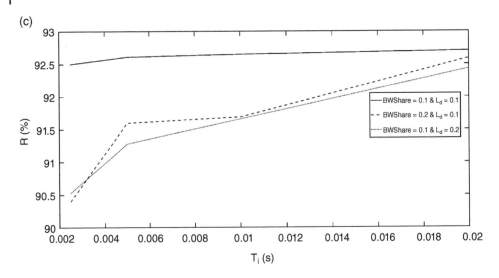

(c)

Figure 3.18 (*Continued*)

Table 3.8 Reliability of the integrated system for different network configurations [BWShare, T_i, L_d] of the two architectures.

Configurations	R (Ethernet)	$E[J(\vec{x})]$ (Ethernet)
[0.2, 0.005, 5%]	92.89%	$[463.51, 4.55 \times 10^4]$
[0.4, 0.005, 5%]	90.43%	$[593.96, 3.80 \times 10^4]$
[0.05, 0.005, 5%]	93.58%	$[433.91, 3.39 \times 10^4]$
[0.2, 0.0025, 5%]	92.62%	$[473.08, 4.40 \times 10^4]$
[0.2, 0.02, 5%]	93.08%	$[433.92, 4.93 \times 10^4]$
[0.2, 0.005, 20%]	90.60%	$[574.40, 6.27 \times 10^4]$
[0.2, 0.005, 1%]	93.09%	$[432.49, 3.55 \times 10^4]$

Configurations	R (Hybrid)	$E[J(\vec{x})]$ (Hybrid)
[0.2, 0.005, 5%]	92.65%	$[471.63, 3.01 \times 10^4]$
[0.4, 0.005, 5%]	90.23%	$[559.57, 4.33 \times 10^4]$
[0.05, 0.005, 5%]	93.43%	$[429.05, 3.25 \times 10^4]$
[0.2, 0.0025, 5%]	92.35%	$[510.06, 3.07 \times 10^4]$
[0.2, 0.02, 5%]	93.05%	$[436.42, 3.00 \times 10^4]$
[0.2, 0.005, 20%]	90.09%	$[623.37, 4.50 \times 10^4]$
[0.2, 0.005, 1%]	92.97%	$[433.17, 4.89 \times 10^4]$

3.4.3 Optimal AGC

3.4.3.1 Optimal Controller Performance
The MOPSO is employed to design the optimal PID controller. In the MOPSO, the number of particles and the number of iterations are set to 50 and 50. For each particle, the stochastic objective function $J(\vec{x}) = [E[J_1(\vec{x})], E[J_2(\vec{x})]]^T$ is estimated from 200 MCS samples.

Table 3.9 lists the optimal control parameters $[K_P, K_I, K_D]$ for the integrated system with three network architectures, which minimize $J(\vec{x})$ under the configuration [0.2, 0.005, 5%]. The reason behind selecting this specific configuration is that the generated time delays match typical time delays in most network conditions, which can be in the order of 100-300 ms [95].

Table 3.9 quantifies the amount of system reliability and control effectiveness that is lost due to the effects of communication degradation with respect to the perfect communication

Table 3.9 Optimal PID controllers $[K_P, K_I, K_D]$ for different network configurations.

Network	Tradeoff Solution	Reliability	$E[J_1(\vec{x})]$
Ethernet	Best Tradeoff I	97.27%	186.30
[0.2, 0.005, 5%]	Best Tradeoff II	97.27%	186.30
	Minimum $E[J_1(\vec{x})]$	97.78%	143.11
	Minimum $E[J_2(\vec{x})]$	12.26%	8.25×10^4
Hybrid	Best Tradeoff I	97.10%	197.82
[0.2, 0.005, 5%]	Best Tradeoff II	97.10%	197.82
	Minimum $E[J_1(\vec{x})]$	97.60%	176.06
	Minimum $E[J_2(\vec{x})]$	12.40%	8.04×10^4
Perfect	Best Tradeoff I	97.79%	160.46
Communication	Best Tradeoff II	97.79%	160.46
	Minimum $E[J_1(\vec{x})]$	97.99%	140.19
	Minimum $E[J_2(\vec{x})]$	12.13%	8.23×10^4

Network	Tradeoff Solution	$E[J_2(\vec{x})]$	Optimal PID
Ethernet	Best Tradeoff I	402.85	$[1.12, 0.094, 0.35]$
[0.2, 0.005, 5%]	Best Tradeoff II	402.85	$[1.12, 0.094, 0.35]$
	Minimum $E[J_1(\vec{x})]$	8.34×10^4	$[7.6, 0, 0.51]$
	Minimum $E[J_2(\vec{x})]$	0	$[0, 0, 0]$
Hybrid	Best Tradeoff I	433.37	$[1.07\ 0.105, 0.41]$
[0.2, 0.005, 5%]	Best Tradeoff II	433.37	$[1.07\ 0.105, 0.41]$
	Minimum $E[J_1(\vec{x})]$	9.84×10^3	$[5, 0.34, 2.02]$
	Minimum $E[J_2(\vec{x})]$	0	$[0, 0, 0]$
Perfect	Best Tradeoff I	559.30	$[1.45, 0.12, 0.4]$
Communication	Best Tradeoff II	559.30	$[1.45, 0.12, 0.4]$
	Minimum $E[J_1(\vec{x})]$	2.19×10^3	$[3.8, 0.04, 0.45]$
	Minimum $E[J_2(\vec{x})]$	0	$[0, 0, 0]$

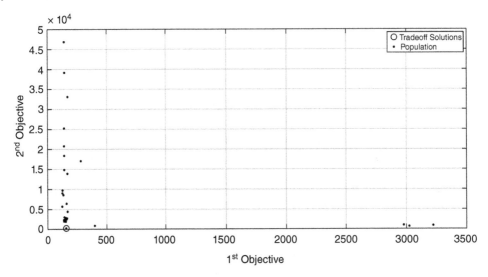

Figure 3.19 Pareto optimal sets and tradeoff of the optimization objectives for the optimal PID controller and perfect communication network.

case. Above results also show that the two objectives are non-commensurable and conflicting with each other. Higher system reliability means more control efforts and vice versa. Therefore, the MOPSO algorithm is better for the optimization of PID controller, rather than combining them. The efficiency of the MOPSO algorithm in finding better combinations with higher reliability index is higher than that of the single-objective PSO, which has been proved by work [273].

As a sample, Pareto optimal set and best tradeoff solutions for the optimal PID controller under perfect communication are been depicted in Figure 3.19. The best compromises, i.e., the Best Tradeoff I and Best Tradeoff II, for the system frequency deviations and control efforts are the same. Results indicate that the reduction in the system frequency deviations leads to the increase in the control efforts, and vice versa. The Pareto front gathers to a very small area toward the maximum iterations.

3.4.3.2 Scenario 1 Analysis

The scenario 1 has low RERs and low load with normal operation and step load change. Figure 3.20 illustrates power-sharing curves of microgrids with three network architectures, i.e., the perfect communication network, Ethernet and hybrid network, when it supplies low load given low RERs in the autonomous mode. The three network architectures are equipped with tradeoff PID controllers for the communication configuration [0.2, 0.005, 5%], given in Table 3.9.

In this scenario, the load is close to but a bit larger than the output of RERs. For the optimum use of energy, the DEG can be shut down (Figure 3.20 (c)), and the deficit power in the microgrid is supplied by the BESS and FESS, shown by Figure 3.20 (a) and Figure 3.20 (b).

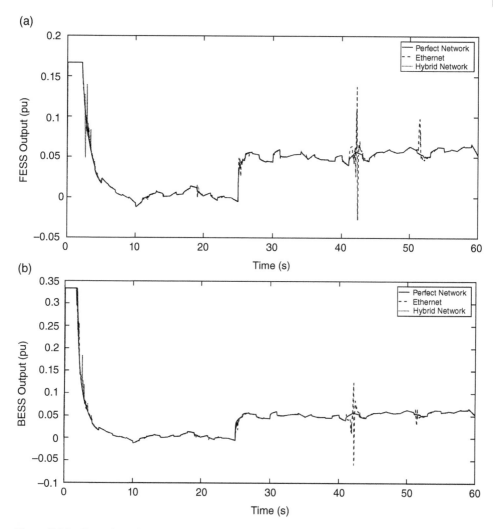

Figure 3.20 Evolution of (a) FESS, (b) BESS, (c) DEG, and (d) system frequency Δf under normal operation and step load change, given low RERs and low load.

Then, the microgrid is influenced by a sudden load stepping increase 0.1 pu at the moment $t = 25$ s, so that the amount of 0.05 pu to FESS and 0.05 pu to BESS are added, illustrated by Figure 3.20 (a) and Figure 3.20 (b). Figure 3.20 (d) shows that the PID controller in the perfect communication case provides the fastest response against the sudden load increase and achieves minimum system frequency deviations.

For the integrated system with Ethernet and hybrid network, the FESS, BESS and DEG output and Δf converge to the same values as the AGC with perfect communication, with small convergence rates due to time delays and packet dropouts. The strong effects of communication degradation introduced by the 802.11b/g in the hybrid network yield

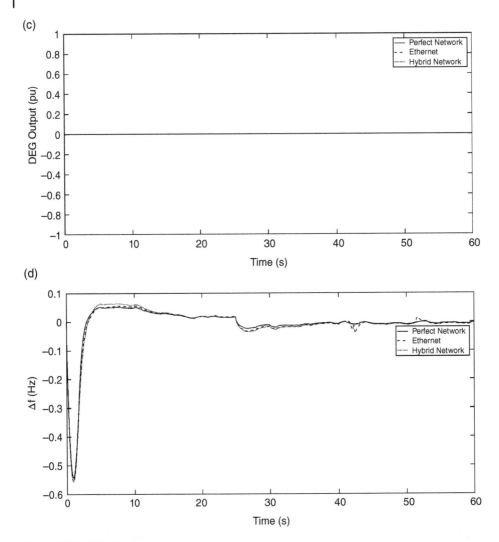

Figure 3.20 (*Continued*)

the smallest convergence rate, compared to the Ethernet. Shown by Figure 3.20 (a), Figure 3.20 (b) and Figure 3.20 (d), the tradeoff PID controllers are effective in eliminating the oscillations in the output of FESS and BESS, and system frequency, caused by random communication degradation.

3.4.3.3 Scenario 2 Analysis
The scenario 2 has low RERs and high load with normal operation and step load change. In this scenario, the load is much larger than the output of RERs, which means that the deficit power cannot be only supplied by the discharging BESS and FESS. The DEG should be connected to the microgrid to help provide complementary power, shown by Figure 3.21 (c).

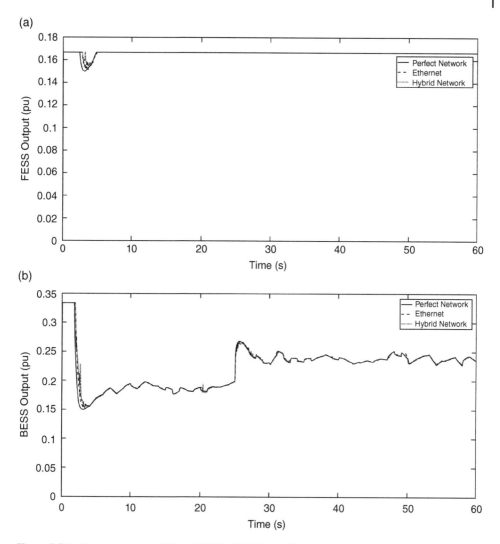

Figure 3.21 Evolution of (a) FESS, (b) BESS, (c) DEG, and (d) system frequency Δf under normal operation and step load change, given low RERs and high load.

Figure 3.21 (a) highlights impacts of time delays in the response of the controllable BESS to the control signal, for the integrated systems with imperfect communication. The outputs of FESS in the Ethernet and the hybrid architecture do not change in the time interval [2.40s, 2.80s], as compared to the perfect communication case. FESS keeps releasing the same power to the microgrid even in unbalanced conditions because the data packets containing the control signals do not reach the destination.

At the same time, the control center lacks enough frequency measurements from the PMU, and issues inaccurate commands. Delays in component response and missing control signals ultimately cause the FESS, BESS and DEG to perform untimely and inaccurately, causing large frequency deviation in Figure 3.21 (d).

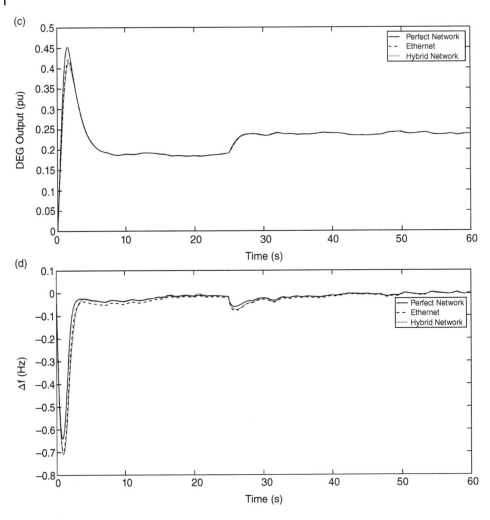

Figure 3.21 (*Continued*)

Then, the microgrid is influenced by a sudden load stepping increase 0.1 pu at the moment $t = 25$ s, so that the amount of 0.04 pu to BESS and 0.06 pu to DEG are added, illustrated by Figure 3.21 (b) and Figure 3.21 (c). As the hybrid network easily causes larger time delays and more packet dropouts, the system frequency deviations of the microgrid with it has the slowest convergence rate to final state ($\Delta f = 0$), compared to other two cases in the Figure 3.21 (d).

3.4.3.4 Scenario 3 Analysis

The scenario 3 has high RERS and low load with normal operation and step load change. In this scenario, the load is smaller than the output of RERs, and therefore the surplus power

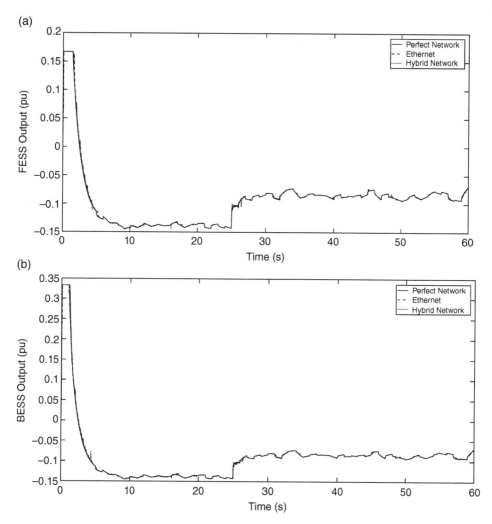

Figure 3.22 Evolution of (a) FESS, (b) BESS, (c) DEG, and (d) system frequency Δf under normal operation and step load change, given high RERs and low load.

should be used to charging the BESS and FESS, shown by Figure 3.22 (a) and Figure 3.22 (b).

In addition, the DEG should be disconnect as shown in Figure 3.22 (c), from the microgrid for the optimum use of energy. When the microgrid is influenced by a sudden load stepping increase 0.1 pu at the moment $t = 25$ s, the charging rates of FESS and BESS are reduced by 0.048 pu and 0.052 pu, respectively. Above power management behaviors are to supply the sudden load increase.

Figure 3.22 (d) indicates that the microgrid with Ethernet is under serious network congestions in the early stage, and therefore larger system frequency deviations and slow convergence rate can be observed. However, above performance indexes are almost as good as

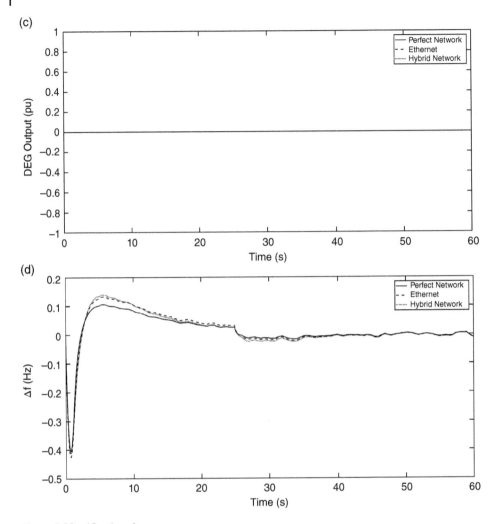

Figure 3.22 (*Continued*)

the perfect communication case because the network conditions of Ethernet become better in the later stage.

3.4.3.5 Scenario 4 Analysis
The scenario 4 has high RERS and high load with normal operation and step load change. In this scenario, both the load and the output of RERs are high. The DEG is still shut down to make most use of RERs, shown by Figure 3.23 (c).

When in the normal operation, both the FESS and BESS are charging shown by Figure 3.23 (a) and Figure 3.23 (b), because the output of RERs is slightly larger than the

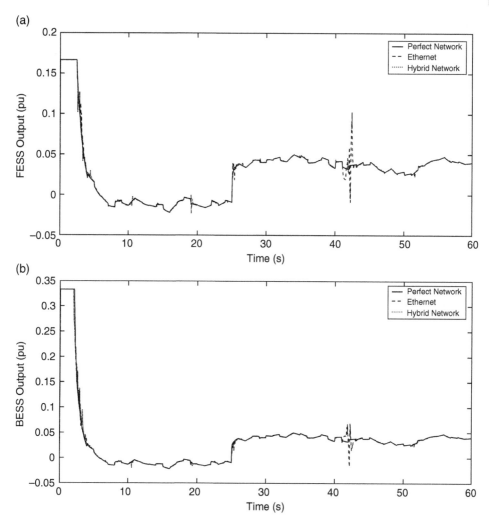

Figure 3.23 Evolution of (a) FESS, (b) BESS, (c) DEG, and (d) system frequency Δf under normal operation and step load change, given high RERs and high load.

load. In addition, in this random scenario, the network condition of hybrid network is worse than that of Ethernet, and therefore the response in the system frequency curve of the microgrid with the Ethernet are slower than that of the microgrid with hybrid network in Figure 3.23 (d).

When the microgrid is influenced by a sudden load stepping increase 0.1 pu at the moment $t = 25$ s, the FESS and BESS are transferred from the charging state to the discharging state, to supply the sudden load increase. Both the network conditions of the Ethernet and hybrid network become well, and therefore the system frequency deviations

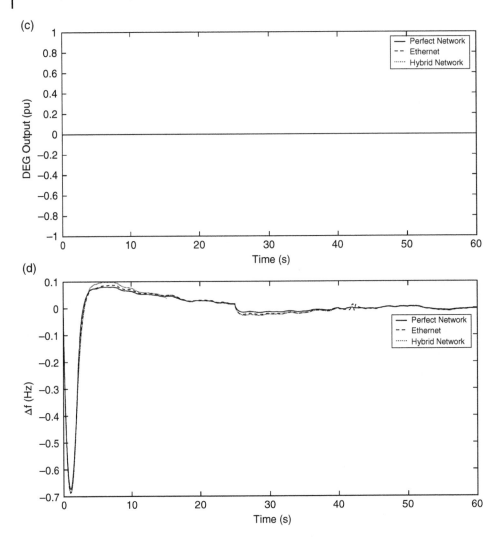

Figure 3.23 (*Continued*)

and convergence rate are close to these of the perfect communication case, illustrated by Figure 3.23 (d).

3.4.3.6 Robustness of Optimal AGC

Finally, this chapter investigate the robustness of the PID-controlled AGC operations optimized for the configuration [0.2, 0.005, 5%] in Table 3.9, subject to increasing communication degradation. Figure 3.24 demonstrates the system frequency deviation Δf in the microgrids with three different network architectures of one MC simulation. Four configurations are investigated, respectively, BWShare increased from 0.2 to 0.4, T_i decreased from 0.005 s to 0.002 s, and L_d increased from 5% to 30%. Similar to section 3.5.4 to 3.5.7,

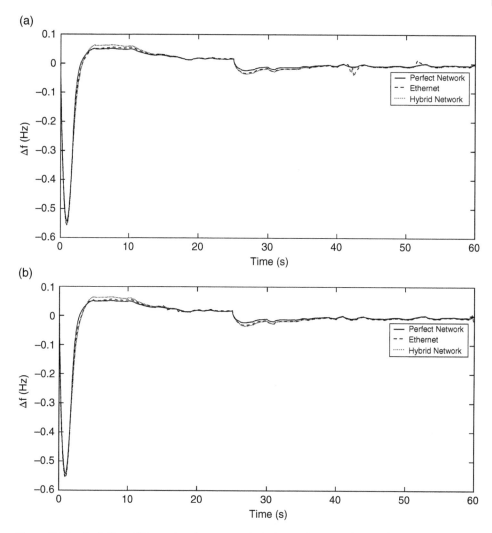

Figure 3.24 Evolution of the system frequency deviation under normal operation and step load change subject to increasing communication degradations (a) [0.2, 0.005, 5%], (b) [0.4, 0.005, 5%], (c) [0.2, 0.002, 5%], and (d) [0.2, 0.005, 30%].

the microgrid is influenced by a sudden load stepping increase 0.1 pu at the moment $t = 25$ s.

Figure 3.24 unveils the abilities of various controllers in providing satisfied AGC performance for unknown network configurations. The optimized controllers for Ethernet and hybrid network are both robust to degraded communication but small frequency fluctuations appear. The reliability in the four Ethernet configurations is, respectively, 97.27%, 97.21%, 97.10% and 96.25%. The reliability in the four hybrid network configurations is, respectively, 97.10%, 97.05%, 96.81% and 93.96%. The analysis of the PID tuning indicates

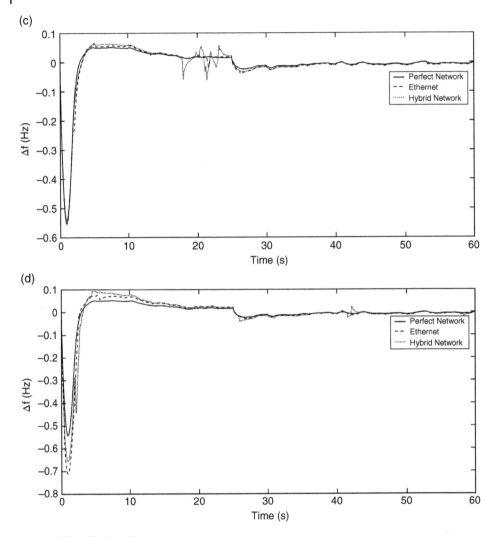

Figure 3.24 (*Continued*)

that the MOPSO produces different optimal PID controllers for different runs of the optimization algorithm [271, 272, 274, 275, 290]. Nonetheless, even if PID controllers tuned using heuristics have different values of R and $E[J(\vec{x})]$, they are capable of stabilizing the system frequency quickly in the analyzed network configurations [72, 74–76, 78, 79].

4

Reliability Analysis of CPSs

Because in many industrial Cyber-Physical Systems (CPSs), the traditional reliability concept, i.e., failure probability, and the control stability are not most critical in terms of stakeholder concern and not easily obtained, this chapter interprets the unsatisfied demand and operation cost of CPSs as the new reliability concept.

4.1 Conceptual DGSs

Distributed Generation Systems (DGS) is one of common CPSs and is typically monitored and controlled by Supervisory Control and Data Acquisition (SCADA) [121]. We consider the interactions between distributed subsystems and centralized control centers (CCC) via degraded communication networks, from a system-of-systems perspective [146, 310].

Figure 4.1 illustrates the conceptual diagram of the integrated system studied in this chapter. It should be noted that data exchanges between the CCC and the power sources in the DGS are bidirectional [142, 143].

4.2 Mathematical Model of Degraded Network

This section illustrates the general models of communication networks with single-packet transmission. The single-packet transmission lumps data into a single packet and transmits it to the destination at one time [140]. Standards of such communication networks are generally set by protocols and most of them involve the interconnections of communication lines to the CCC via Local Area Networks (LAN), Home Area Networks (HAN) and Wide Area Networks (WAN) [143].

This type of transmission is suitable for communication networks with large packet size, such as Ethernet (v2) with a Maximum Transmission Unit (MTU) of 1500 Byte (B), LAN (Token Ring, 802.5) with an MTU of 4464 B, and Wireless Local Area Networks (WLAN) (802.11) with an MTU of 7981 B. They are widely used for communications between power sources and CCC in DGSs [140, 311, 312].

Part of the contents are from [89] and permission has been obtained to use the contents.

Cyber-Physical Distributed Systems: Modeling, Reliability Analysis and Applications, First Edition.
Huadong Mo, Giovanni Sansavini and Min Xie.
© 2021 John Wiley & Sons Ltd. Published 2021 by John Wiley & Sons Ltd.

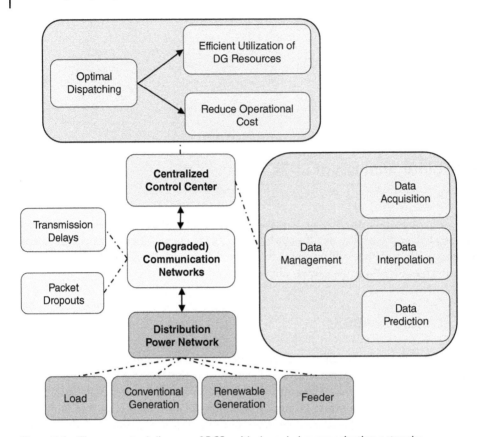

Figure 4.1 The conceptual diagram of DGSs with degraded communication networks.

In reality, the transmission delays and packet dropouts are most common and usually hinder the performance of the integrated system. Their stochastic behaviors have been studied in many works [313, 314].

4.2.1 Model of Transmission Delay

The continuous transmission delay τ_k takes place at each packet transmission and is inevitable, where k means the kth period of data exchange. Transmission delay is usually bounded [148, 149, 313, 315, 316]:

$$\tau^{lb} \leq \tau_k \leq \tau^{max} \leq \tau^{ub} \tag{4.1}$$

where τ^{lb} is the lower bound of the transmission delay, which is larger than zero and usually equals the transmission time in the physical medium, τ^{ub} is the upper bound and τ^{max} is the maximum allowable delay time (MADT). If the transmission time exceeds this value, the data packet is often regarded as missing [148, 313].

In real applications, transmission delay can be easily measured using the time-stamp method. τ_k is described by a Markov process-reflected Wiener process, which takes value within the predefined bounds in (4.1). Based on the principle-symmetry reflection principle of the Wiener process, if the path of $w(t)$ reaches a bound B^+ (τ^{lb} or τ^{ub}) at time $t = t_r$, then the subsequent path after time t_r has the same distribution as the reflection of the path above the bound [147].

The new process is defined in a corollary of the strong Markov property of Brownian motion, as follows:

$$\widetilde{w}(t) = \begin{cases} w(t), & \text{for } t \le t_r \\ 2B^+ - w(t), & \text{for } t > t_r \end{cases} \tag{4.2}$$

where $w(t)$ is the standard Wiener process, which has increments $w(t) - w(s) \sim a\sqrt{t-s}N(0,1)$ for $0 \le s < t$ and a is the power coefficient. In this chapter, the times s and t can only take value as kT, where T is the communication interval.

Multiple reflections may exist if the increment is much larger than the interval between the lower and upper bounds. This case is also taken into account and (4.2) can be modified as follows:

For $t \le t_r$,

$$\widetilde{w}(t) = w(t)$$

For $t > t_r$,

$$\widetilde{w}(t) = (-1)^v[I(v)(v+2) + (1 - I(v))(v+1)]B^+ \tag{4.3}$$

$$-(-1)^v[I(v)v + (1 - I(v))(v+1)]B^- - (-1)^v w(t)$$

$$I(v) = \begin{cases} 1, & \text{if } v \text{ is an even number} \\ 0, & \text{otherwise} \end{cases}$$

$$v = \lfloor |(w(t) - B^+)/(B^+ - B^-)| \rfloor$$

where $\lfloor x \rfloor$ stands for the largest previous integer to x.

4.2.2 Model of Packet Dropout

This section describes the general model of packet dropout. Let $x^{(0)}$ and $\overline{x}^{(0)}$ denote the data packet sent by the local power source and the data packet used by CCC to compute Optimal Power Flow (OPF), respectively.

In the period k, one power source of DGSs to CCC transmits the data packet $x_{k-1}^{(0)}$ containing the power data measured at period $k-1$ through the communication networks. The buffer of CCC is sufficiently large to store all received data packets.

If there is no packet dropout, the buffer can receive the new data packet and, then, CCC picks it out as $\overline{x}_{k-1}^{(0)}$ to compute the OPF for period k; otherwise, CCC has to use the predicted power data as $\overline{x}_{k-1}^{(0)}$. Thus:

$$\overline{x}_{k-1}^{(0)} = \begin{cases} x_{k-1}^{(0)}, & \text{if } N_k = 0 \\ PM(\overline{x}^{(0)}), & \text{otherwise} \end{cases} \tag{4.4}$$

where N_k is the quantity of packet dropouts at period k, which is counted from the current period k to the last successful packet transmission at period $k - N_k$. PM is the prediction model, which will be presented in Section 4.4.1.

It should be noticed that N_k is generally bounded [140, 317, 318]:

$$0 \le N_k \le N^{max} \tag{4.5}$$

where N^{max}, a nonnegative integer, is the number of maximum allowable packet dropouts.

The packet dropout can be detected by checking whether the buffer can receive data packets before the maximal allowable transmission delay. In order to relieve the assumption that packet dropout in each period is independent from the others, it is modeled by a multi-state Markov chain [140, 317].

The Markov chain takes value from $N = \{0, 1, ..., N^{max}\}$, with the transition probability matrix $P = [\lambda_{ij}]$. The transition probability of jumping from mode i to mode j is defined as

$$\lambda_{ij} = \Pr\left(N_{k+1} = j \mid N_k = i\right) \tag{4.6}$$

where $\lambda_{ij} \ge 0$, $i, j \in N$ and $\sum_j \lambda_{ij} = 1$. According to the definition, the transition probability should satisfy

$$\lambda_{ij} = 0 \quad \text{if} \quad j \ne i+1 \quad \text{and} \quad j \ne 0 \tag{4.7}$$

The homogenous Markov chain with constant transition probability matrix is only adequate for the case where the information on the stochastic behavior of packet dropout is completely known. However, in practice due to limited observations, this information is often partially known.

The polytopic method is applied to describe the partially known transition probability [148, 317–319]. Let P_i be the ith row of P, which is partially known but belongs to a convex set with the known vertices $P_i^{s_i}$:

$$P_i \in \left\{ \sum_{j=1}^{s_i} \alpha_j P_i^j, \ \sum_{j=1}^{s_i} \alpha_j = 1, \ \alpha_j \ge 0 \right\} \tag{4.8}$$

where P_i^j ($j = 1, 2, \cdots, s_i$) are the vertices of P_i indicating the polytope of the ith row, α_j is the coefficient denoting the weights of the vertices to determine P_i and s_i is the total number of the vertices in the ith row, which is determined by the number of unknown elements, see *Remark 4.1* below for details.

Example 4.1 Consider the following matrix

$$P = \begin{pmatrix} ? & ? & 0 \\ ? & 0 & 0.4 \\ 1 & 0 & 0 \end{pmatrix}$$

One reasonable choice for the vertex set for each matrix row is shown as follows:

$P_1^1 = [1\ 0\ 0], \ P_1^2 = [0\ 1\ 0], \ s_1 = 2;$

$P_2^1 = [0.6\ 0\ 0.4], \ s_2 = 1;$

$P_3^1 = [1\ 0\ 0], \ s_3 = 1.$

Based on (4.8), P_i for each row can be written as:

$$P_1 = [?\ 0\ ?] = \alpha_1 P_1^1 + \alpha_2 P_1^2, \ \sum_{i=1}^{2} \alpha_i = 1, \alpha_i \ge 0, i = 1, 2;$$

$$P_2 = [0.6\ 0\ 0.4] = P_2^1;$$

$$P_3 = [1\ 0\ 0] = P_3^1.$$

Remark 4.1   As discussed above, if a row has none or only one "?", based on the normalization constraint, only one vertex needs to be determined. For a row containing more than one "?", the same number of vertices needs to be determined.

The reason for adopting this convex combination method is that the corresponding parameters can be easily estimated, and optimization can be conducted using convex optimization approaches [320]. One popular method to estimate the unknown parameters is based on MLE, conditional on the starting state of the Markov chain [321].

4.2.3   Scenarios of Degraded Network

The simulation results are presented in Figure 4.2 and Figure 4.3 for different simulated days. The communication interval is set as 5 minutes, in accordance with the 5-minute dispatch interval.

The dotted line is the maximum allowable transmission delay time. It can be clearly observed that the delayed data packet with the transmission time exceeding the threshold is regarded as one dropout, as indicated by (4.1). The simulated degraded communication networks are the superposition of transmission delay and packet dropout and resemble the real communication networks.

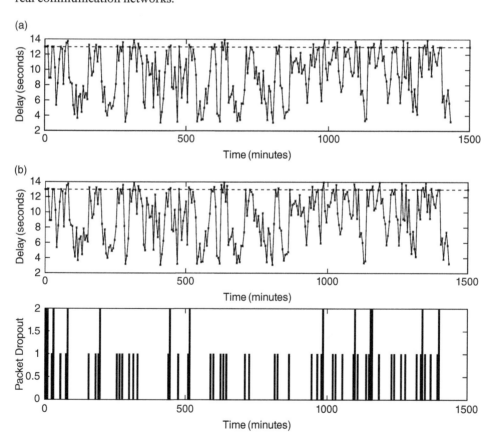

Figure 4.2   Simulation of degraded communication networks - (a) delays and (b) packet dropout for September 14, 2015.

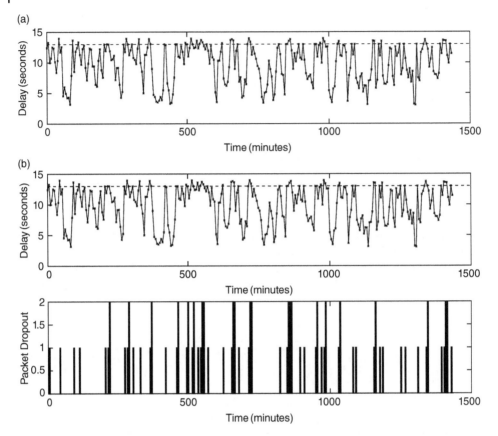

Figure 4.3 Simulation of degraded communication networks - (a) delays and (b) packet dropout for September 21, 2015.

4.3 Modeling and Simulation of DGSs

4.3.1 DGS Model

4.3.1.1 Preliminary Model

The DGS contains main supply (conventional generation, Main Supply (MS)), Distributed Generation (DG) technologies (Photovoltaic Power (PV) and Wind Turbine Generator (WTG)), Electrical Vehicles (EVs), and storage devices (STs), nodes and feeders (FDs). The nodes are regarded as fixed spatial locations at which DGs, and loads can be deployed. Feeders connect different nodes and the power is distributed by them.

Following [322], the configuration of different power sources deployed in DGSs is represented by the following matrix:

$$
\Xi =
\begin{pmatrix}
MS_1 & \cdots & MS_j & \cdots & MS_m & DG_1 & \cdots & DG_j & \cdots & DG_d \\
\xi_{1,1} & \cdots & \xi_{1,j} & \cdots & \xi_{1,m} & \xi_{1,m+1} & \cdots & \xi_{1,m+j} & \cdots & \xi_{1,m+d} \\
\vdots & \ddots & \vdots & \ddots & \vdots & \vdots & \ddots & \vdots & \ddots & \vdots \\
\xi_{i,1} & \cdots & \xi_{i,j} & \cdots & \xi_{i,m} & \xi_{i,m+1} & \cdots & \xi_{i,m+j} & \cdots & \xi_{i,m+d} \\
\vdots & \ddots & \vdots & \ddots & \vdots & \vdots & \ddots & \vdots & \ddots & \vdots \\
\xi_{n,1} & \cdots & \xi_{n,j} & \cdots & \xi_{n,m} & \xi_{n,m+1} & \cdots & \xi_{n,m+j} & \cdots & \xi_{n,m+d}
\end{pmatrix}
\tag{4.9}
$$

where MS_j denotes the type of MS, DG_j denotes the type of DG, n is the number of nodes in the DGS, m is the number of main supply types, d is the number of DG technologies and $\xi_{i,j}$ is the number of units of the power source type j deployed at node i.

Locations of feeders are described by the set of pairs of connected nodes:

$$FD = \{(1,2), \cdots, (i, i')\}, \forall (i, i') \in N \times N \tag{4.10}$$

where (i, i') denotes a feeder in the DGS.

The degradation processes of DGs, MSs and feeders are also taken into account. They can cause unexpected events, such as interruption of the DG power outputs and reduction of the controllability of the power flow, which would affect components availability and eventually increase the operational cost.

For simplicity, the uncertain behavior of components' failures and corresponding repair activities is modeled by a continuous-time Markov chain (CTMC). The components have two mutually exclusive states: operating and under repair (completely failed) [322–324]. The components availability is determined by the stochastic process of transition between under repair and working states which is modeled by Markov chain.

Given the assumption that the duration of each state follows an exponential distribution, the mechanical state evolutions of a component can be obtained:

$$mc = \begin{cases} 1, & \text{if the component is operating} \\ 0, & \text{otherwise} \end{cases} \tag{4.11}$$

$$f_{mc}(mc) = \frac{(1 - mc)\lambda^F + mc\lambda^R}{\lambda^F + \lambda^R}$$

where the power source belongs to matrix $\{\Xi, FD\}$, mc is the binary mechanical state variable and belongs to set $\{0, 1\}$, λ^F and λ^R are the failure rate and repair rate (/h), respectively, and $f_{mc}(mc)$ is the mechanical state probability mass function.

4.3.1.2 Power Source Model

Single load profiles, MS profiles, PV profiles, WTG profiles and EV profiles in the nodes of the DGS are generated as 5-minute interval power curves. As shown in Section 4.3, these profiles can be interpolated from raw power data, e.g., GRID DATA in the period from January 1, 2015 to September 22, 2015 provided by Elia, Belgium [325]. The model based on the interpolated power data is continuous and contains the uncertainties in the raw power data.

Figure 4.4 depicts the single load, MS, PV and WTG profiles on September 22, 2015. EVs are regarded as plug-in hybrid electric vehicles (PHEVs) with three possible operating states: charging, discharging and disconnected. When considering the performance of the EV ensemble, various PHEV charging load profiles (PCLPs) can be used [314, 326]. Note that the PCLP depends more on the movements of the vehicles rather than those of the humans: we, then, extract PCLP from the data of vehicle trips instead of personnel trips [327, 328].

The PHEVs are characterized by their all-electric ranges (AERs) and the charging times are determined by their charging levels. For simplicity, the total required electric power for PHEV is discretized into finite numbers of charging levels with step size of 1.4 kW per hour [314]. Two charging strategies are considered: power scaling approach, where the charging time is constant and the delivered power during each minute is scaled; time scaling approach, where PHEVs are charged by the maximum available power while charging time is flexible.

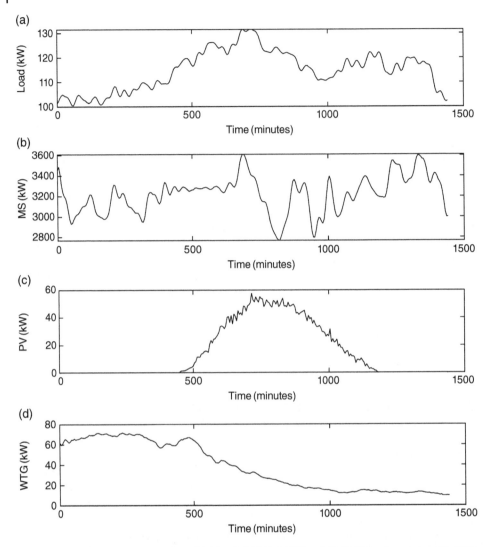

Figure 4.4 Different power profiles of (a) load, (b) MS, (c) PV and (d) WTG for September 22, 2015.

Figure 4.5 demonstrates different PCLPs (sum of 30 individual PCLPs) based on different AERs and charging schedules (20, 30 and 40 miles) for September 14, 2015.

In this model, the STs are mainly the batteries with two operating states: charging and discharging. Different from some previous works which model the random behaviors of batteries as time-independent [322, 329], a continuous state evolution model is adopted for more accurate representation of the real-world situations.

At a certain time period k, state of charge (SOC) is used to quantify the amount of available energy in the battery [330, 331]. It is defined as

$$SOC_k = \frac{Q_k}{Q_n} \tag{4.12}$$

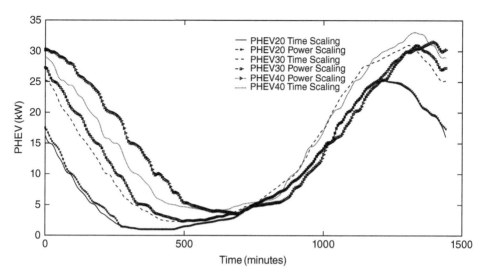

Figure 4.5 Different PCLPs for September 14, 2015.

where Q_k[kWh] is the real-time energy of the battery at period k and Q_n[kWh] is the nominal energy of the battery when fully charged.

Note that when CCC calculates OPF, the following constraints for real power flow P_k^{ST} of the batteries at period k should be satisfied [330, 332]:

When batteries are discharging

$$P_k^{ST} \cdot T \le \min\{SOC_k \cdot Q_n, P_R^{ST} \cdot T\} \tag{4.13}$$

When batteries are charging

$$P_k^{ST} \cdot T \le \min\{(1 - SOC_k) \cdot Q_n, P_R^{ST} \cdot T\} \tag{4.14}$$

where P_R^{ST} is the rated power [kW].

Therefore, SOC_{k+1} is determined as

$$SOC_{k+1} = \frac{Q_k + (-1)^I \cdot P_k^{ST} \cdot T}{Q_n} = SOC_k + (-1)^I \cdot \frac{P_k^{ST} \cdot T}{Q_n} \tag{4.15}$$

where $I = \begin{cases} 1, & \text{when discharging} \\ 0, & \text{when charging} \end{cases}$.

Clearly, this model enables CCC to continuously monitor and control the storage devices.

4.3.2 Data Interpolation

The details can be referred to the NREL report [333]. The raw power data used in our model has been recorded at 15-minute intervals from GRID DATA. For the real-time data exchange and OPF calculation, the integrated system requires data sampled at a time interval of 5 minutes. The 15-minute to 5-minute conversion can be done through the Fourier transform methods [334].

The main procedures of the Fast Fourier Transform (FFT) method [334–336] are as follows: 1) apply a cubic spline interpolation to the 15-minute data to achieve power estimations at 5-minute intervals; 2) use high-frequency magnitude data integrated with randomized phase to generate the noise term of the power data. Real and imaginary terms are computed using magnitude and phase, which result in the high-frequency noise after applying the inverse Fourier transform. Thus, the simulated power data is generated by the superposition of the cubic spline interpolation and this noise.

Figure 4.6 to Figure 4.9 make comparisons between the simulated and the actual data for a short period of time (i.e., 1500 minutes), considering different time intervals and different

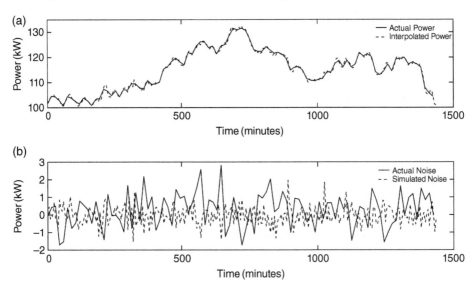

Figure 4.6 Comparison of (a) interpolated 5-minute load data to actual load data and (b) simulated noise to actual noise for September 14, 2015.

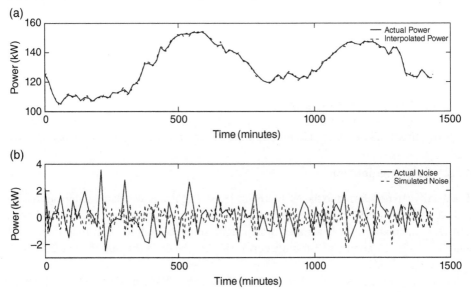

Figure 4.7 Comparison of (a) interpolated 5-minute load data to actual load data and (b) simulated noise to actual noise for September 21, 2015.

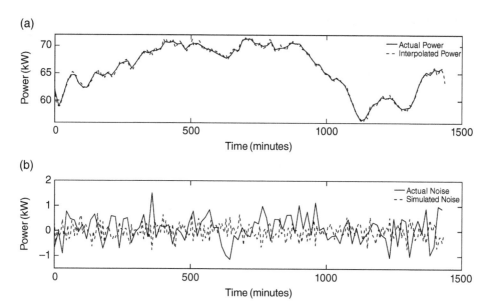

Figure 4.8 Comparison of (a) interpolated 5-minute wind power data to actual data and (b) simulated noise to actual noise for September 14, 2015.

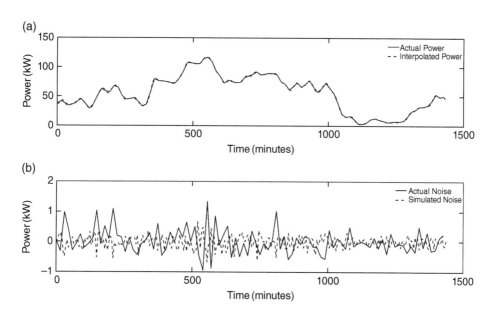

Figure 4.9 Comparison of (a) interpolated 5-minute wind power data to actual data and (b) simulated noise to actual noise for September 22, 2015.

power sources (i.e., load and wind). The actual noise was computed by subtracting the cubic spline fit and simulated noise from the actual power data [334]. Note that the noise of the simulated data is more uniform and smaller in magnitude than that of the actual data. Nevertheless, the simulated data does resemble the actual one.

4.4 Reliability Estimation Via OPF

The operational procedures of the integrated system of systems are follows: 1) power sources send the measured real-time power data to CCC via communication networks; 2) CCC performs necessary data prediction for each power source and determines the optimal power dispatching for the next period using OPF; 3) control signals are sent to the respective power sources through communication networks; 4) each power source adjusts its output according to the control signal if received, otherwise according to the predicted control signal made by itself.

4.4.1 Data Prediction

As mentioned earlier, the need of prediction arises due to the imperfect data exchange induced by the degraded communication networks. CCC performs the prediction for the OPF computation of next period, considering both transmission delay and packet dropout as the model PM in (4.4). Local DGs also perform prediction, if packet dropout occurs. The grey prediction model is widely applied for power data [337, 338], standing between white-box and black-box models to deal with partially uncertain factors.

The basic of the grey prediction model is based on differential modeling. Grey differential model (GM) is a dynamic model mathematically described by a system of differential equations. Some issues need to be accounted for [339]: a grey process refers to a stochastic process whose amplitudes are a function of time; grey modelling is determined by the generating series rather than the raw one; to develop a GM, corresponding grey derivative, parallel shooting and differential equations need to be defined; to develop a GM, few data points (as few as 4) are needed. The grey model used in this chapter is the GM (1, 1), whose details can be found in the following example.

Example 4.2   Let $x^{(0)}$ be the series of the raw power data. Given $x^{(1)} = \sum_{m=1}^{k} x^{(0)}(m)$, a GM is defined as

$$x^{(0)}(k) + az^{(1)}(k) = b, k = 1, 2, \cdots, n, \cdots$$

where $z^{(1)}(k) = 0.5x^{(1)}(k) + 0.5x^{(1)}(k - 1)$, a and b represent the developing coefficient and the grey input, respectively, and $x^{(0)}(k)$ is a grey derivative that maximizes the information density for a given power data.

By least squares method, a and b can be obtained

$$\hat{a} = \left(\frac{a}{b} \right) = (B^T B)^{-1} B^T y_N$$

where B is a data matrix and y_N are given by

$$B = \begin{pmatrix} -z^{(1)}(2) & 1 \\ \cdots & \cdots \\ -z^{(1)}(n) & 1 \end{pmatrix} \text{ and } y_N = \begin{pmatrix} x^{(0)}(2) \\ \vdots \\ x^{(0)}(n) \end{pmatrix}$$

A shadow for $x^{(0)}(k) + az^{(1)}(k) = b$ can be determined as

$$\frac{dx^{(1)}}{dt} + a \otimes (x^{(0)}) = b$$

Then, the following equation can be derived

$$\mathfrak{B}^{(1)} = \otimes(x^{(1)}) = \{x^{(1)} \mid \forall x^{(1)} \in \mathfrak{B}^{(1)} = \{\widetilde{\otimes}(x^{(1)}) | \mathfrak{L}\}$$

where $\otimes(x^{(1)})$ is a background grey number of $dx^{(1)}/dt$, $\widetilde{\otimes}(x^{(1)})$ is the whitening value of the grey number of $\otimes(x^{(1)})$ and $\mathfrak{L}$ stands for the parallel morphism or shooting from $\widetilde{\otimes}(x^{(1)})$ to the components of $dx^{(1)}/dt$ with the following form

$$\mathfrak{L} \begin{array}{l} \nearrow x^{(1)}(t + \Delta t)/\Delta t \\ \searrow x^{(1)}(t - \Delta t)/\Delta t \end{array}$$

Therefore, the response equations for GM (1, 1) can be derived as

$$\hat{x}^{(1)}(k + 1) = \left(x^{(0)}(1) - \frac{b}{a}\right) e^{-ak} + \frac{b}{a}$$

$$\hat{x}^{(0)}(k + 1) = \hat{x}^{(1)}(k + 1) - \hat{x}^{(1)}(k)$$

where $\hat{x}^{(1)}(k)$ and $\hat{x}^{(0)}(k)$ indicate the computed values of $x^{(1)}$ and $x^{(0)}$ at period k, respectively.

The model GM (1, 1) and its improved versions (FGM (1, 1), FRMGM (1, 1), NGBM (1, 1) etc.) are widely used in realistic applications, in relation to grey forecasting, grey programming and grey control [340–343].

Figure 4.10 presents the predictions of load data for a 5-hour period on September 14, 2015 using GM (1, 1). The node sends the load data to CCC via the communication networks. As indicated by (4.4), if packet dropout occurs, the data stored in CCC will be the same as the predicted data, as shown in Figure 4.10. Then, the predicted data is used for solving OPF for the next period.

4.4.2 MCS of DGSs

There are two main techniques for assessing the performance of the integrated system of systems: analytical methods and Monte Carlo simulation (MCS) [323, 344]. The analytical methods can obtain explicit solutions at the price that the results may not reflect realistic features, like dynamic generation/load profiles and complex system structures, because of the simplifying assumptions needed [322].

MCS is capable of handling more realistic situations by simulation runs of virtual system life and the quantities of interest are estimated from their statistics, thus avoiding solving difficult equations [345]. For these reasons, MCS is suitable for the quantitative analysis of the integrated system of systems taking into account uncertainties [124, 324, 329].

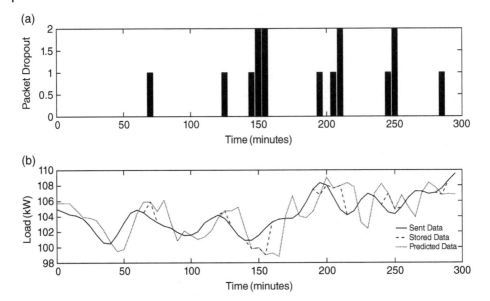

Figure 4.10 Simulation of (a) packet dropout and its impact on (b) load data prediction for September 14, 2015.

The sequential MCS is used to simulate the operations of the integrated system for a certain duration, generating random events considering their time dependence, e.g., simulating the degraded communication networks, randomly sampling loads and distributed power sources, and simulating energy prices and contingencies.

Given the configuration of the DGS $\{\Xi, FD\}$ defined by (4.9) and (4.10), continuous scenarios ϑ of operational conditions are generated. At the kth period, scenario ϑ_k ($\vartheta_k \in \vartheta$) is represented by

$$\vartheta_k = [t_k, \tau_k, N_k, P_{i,j}^{ms}, L_i, PV_i, W_i, EV_i, Q_i, mc_{i,j}, mc_{(i,i')}] \tag{4.16}$$

where $i, i' \in N, j \in MS \cup DG, (i, i') \in FD$, t_k is the date of kth period, τ_k and N_k are the transmission delay and packet dropout, respectively, $P_{i,j}^{ms}, L_i, PV_i, W_i, EV_i$ and Q_i are the power outputs of type j MS, L, PV, WTG, EV and ST at node i, $mc_{i,j}$ is the mechanical state of power source of type j at node i and $mc_{(i,i')}$ is the mechanical state of feeder (i, i'). These variables are updated at each period.

The sampling of the scenario ϑ_k provides the inputs for OPF analysis, which optimally dispatches the available power $P_{Ga}^{\vartheta_k}$ to supply the loads L_i [139–141, 146]. At period k, the available output $P_{Ga_{i,j}}^{\vartheta_k}$ of power source type j at node i is determined by

$$P_{Ga_{i,j}}^{\vartheta_k} = \xi_{i,j} mc_{i,j} G_j(\vartheta_k) \tag{4.17}$$

where $\xi_{i,j}$ is the number of units allocated and G_j is the power output function associated to the power source type j.

4.4.3 OPF of DGSs

The Direct Current (DC) model is considered in our case. It assumes that power losses are negligible and the voltage throughout the entire power grid is a constant value. This results in a linear power flow formulation, achieving simplicity and reducing computational effort [120, 322].

For each scenario ϑ_k ($\vartheta_k \in \vartheta$), the DC-OPF formulation is given as follows:

$$\min \ Co^{\vartheta_k}(Pu, \Delta\delta) = \sum_{i \in N} \sum_{j \in PS} (Cov_j^{PS} - ep^{\vartheta_k})Pu_{i,j} \tag{4.18}$$

$$+ \sum_{(i,i') \in FD} Cov_{i,i'}^{FD}|B_{i,i'}(\delta_i - \delta_{i'})| + (Cop + ep^{\vartheta_k})\sum_{i \in N} LS_i$$

s.t.

$$L_i^{\vartheta_k} - LS_i - \sum_{j \in PS} Pu_{i,j} - \sum_{i \in N} mc_{(i,i')}^{\vartheta_k} B_{i,i'}(\delta_i - \delta_{i'}) = 0 \tag{4.19}$$

$$0 \le Pu_{i,j} \le Pa_{i,j}^{PS;\vartheta_k} \tag{4.20}$$

$$|B_{i,i'}(\delta_i - \delta_{i'})| \le V^{NET}A_{i,i'}^{FD} \tag{4.21}$$

where $i, i' \in N, j \in MS \cup DG, (i, i') \in FD, Co^{\vartheta_k}$ is the global cost associated with scenario ϑ_k, Cov_j^{PS} is the variable operational cost of the power source type j, $Pu_{i,j}$ (kW) is the used power from the power source type j at node i, $Cov_{i,i'}^{FD}$ and $B_{i,i'}(1/\Omega)$ are the variable operational cost and the susceptance of the feeder (i, i'), δ_i is the voltage angle at node i, Cop is the penalty cost for kWh power demand not supplied.

V^{NET} and $A_{i,i'}^{FD}$ are the nominal voltage of the DGS and the ampacity of the feeder (i, i'), respectively, LS_i(kW) is the load shedding, which is the amount of load disconnected at node i to avoid congestions in the feeders and to balance the power demand with the power supply available [329]. All the costs are in \$/kWh.

This chapter adopts the concept of "price taker" in the DGS and assumes a strong correlation between the overall loads of ϑ_k and the energy price ep^{ϑ_k}. Then, the energy price is determined by an intermediate correlation [329, 346]:

$$ep^{\vartheta_k}(OL) = ep_h\left(-0.38\left(\frac{OL(t_k)}{OL_h}\right)^2 + 1.38\frac{OL(t_k)}{OL_h}\right) \tag{4.22}$$

where ep_h is the energy price with respect to the highest value of overall load OL_h. The overall load $OL(t_k)$ at time t_k is the sum of all loads in scenario ϑ_k.

The global cost Co^{ϑ} of a set of scenarios $\vartheta = \{\vartheta_1, \cdots, \vartheta_k, \cdots, \vartheta_K\}$ within the duration equals to $\sum_{k=1}^{K} Co^{\vartheta_k}$, where K is the number of time intervals. Taking into account the fixed investment cost C_i, the global cost (GC) is defined as

$$GC^{\vartheta} = \sum_{k=1}^{K} Co^{\vartheta_k} + C_i \tag{4.23}$$

$$C_i = \frac{1}{TH}\sum_{i \in N}\sum_{j \in DG} \xi_{i,j}ci_j \tag{4.24}$$

where Co^{ϑ_k} is the global cost for each scenario ϑ_k, TH is the lifetime of the DGS and ci_j is the investment cost of power source type j.

MCS-OPF generates *NS* samples and each is associated with a global cost GC^ϑ defined by (4.23) and (4.24). The global cost vector $GC = \{GC_1^\vartheta, \cdots, GC_{NS}^\vartheta\}]$ contains realizations of the probability mass function of GC and statistics can be computed, like the expected global cost EGC.

4.4.4 Actual Cost and Reliability Analysis

In the integrated system, OPF is performed by CCC. To overcome imperfect information exchange due to time delay and packet dropout, CCC may use the predicted data, derived from the stored data, as the input to OPF model.

Differences between actual data and predicted data may obviously arise, e.g., overestimating the power available from the generation units, underestimating the loads, etc. Thus, CCC might not have the completely accurate information about the entire DGS. Then, the optimal power dispatching given by (4.18) to (4.21) could be different from the actual one.

Additionally, the data packet containing the dispatching commands is sent to each power source via the communication networks. There is a probability that the power source may not receive it due to the packet dropout and, thus, the unit has to perform the dispatch based on its own prediction. Due to inadequate information, the power source may eventually generate more/less output and the feeder has a different power flow.

Load shedding and redundant power output appear in the DGS. In these cases, to avoid system collapse, the DGS automatically determines the power purchased from or delivered to the external grid after counting the total load shedding or total redundant power for all nodes [347, 348].

Therefore, the actual global cost could be significantly different from the operational cost obtained by OPF analysis at CCC. The actual global cost ACo^{ϑ_k} of scenario ϑ_k is expressed as

$$ACo^{\vartheta_k}(\overline{Pu}, \Delta\overline{\delta}) = \sum_{i \in N} \sum_{j \in PS} (Cov_j^{PS} - ep^{\vartheta_k})\overline{Pu}_{i,j}$$

$$+ \sum_{(i,i') \in FD} Cov_{i,i'}^{FD} |B_{i,i'}(\overline{\delta}_i - \overline{\delta}_{i'})| + CopR \cdot \psi \left(\sum_{i \in N} RP_i, \sum_{i \in N} \overline{LS}_i \right) \quad (4.25)$$

$$(Cop + ep^{\vartheta_k}) \cdot \psi \left(\sum_{i \in N} \overline{LS}_i, \sum_{i \in N} RP_i \right) + CopD \cdot \min \left(\sum_{i \in N} \overline{LS}_i, \sum_{i \in N} RP_i \right)$$

where $\overline{Pu}_{i,j}$ (kW) is the power actually used at node i from the power source type j, $\overline{\delta}_i$ is the actual voltage angle at node i, $\overline{LS}_i$(kW) is the actual load shedding at node i, $CopR$ is the penalty cost for kWh of redundant power and RP_i is the redundant power at node i, ψ is defined as $\psi(x,y) = \begin{cases} x, & \text{if } x > y \\ 0, & \text{otherwise} \end{cases}$, $CopD$ is the re-distributing cost for the kWh power processed by the automatic protection mechanism of the DGS [337].

The power flow at node i should satisfy the following equation,

$$\overline{L}_i + RP_i - \overline{LS}_i - \sum_{j \in PS} \overline{Pu}_{i,j} - \sum_{i \in N} mc_{i,i'} |B_{i,i'}(\overline{\delta}_i - \overline{\delta}_{i'})| = 0 \quad (4.26)$$

where $\overline{L}_i$ is the actual load at node i. Note that at the same node, the product of RP_i and $\overline{LS}_i$ always equals to zero.

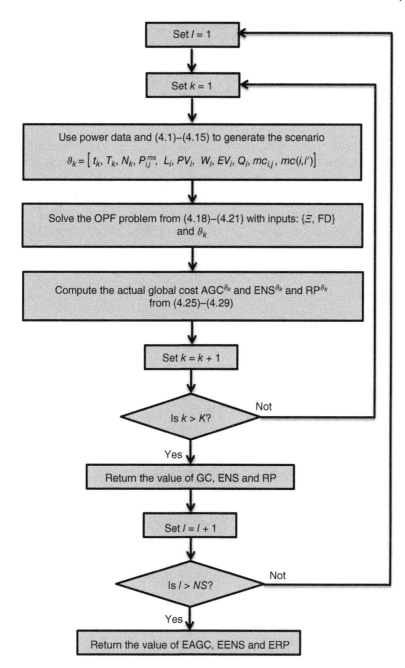

Figure 4.11 Flow chart of MCS-OPF method.

Therefore, the actual global cost of scenarios ϑ is computed as

$$AGC^{\vartheta} = \sum_{k=1}^{NS}(ACo^{\vartheta_k} + C_i) \tag{4.27}$$

In order to study the impact of the degraded communication networks and data prediction on the performance of the DGS, the global costs of the following three cases are computed.

The first case considers the communication networks as perfect, i.e., without transmission delay and packet dropout. The second case considers the imperfect communication networks with data prediction. The third case considers the imperfect communication networks without data prediction.

The global cost of the first ideal case $AGC1^{\vartheta}$ is regarded as a reference value; that of the second case $AGC2^{\vartheta}$ is used to analyze the ability of the data prediction to compensate the influence caused by imperfect communication networks; that of the last case $AGC3^{\vartheta}$ is used to study the overall performance of the integrated system without data prediction.

Energy not supplied (ENS) and redundant power (RP) are common indices for the reliability evaluation of power systems [124, 322, 323]. For scenarios ϑ with K time intervals, the overall ENS and RP are computed as

$$ENS^{\vartheta} = \sum_{k=1}^{K} ENS^{\vartheta_k} \tag{4.28}$$

$$RP^{\vartheta} = \sum_{k=1}^{K} RP^{\vartheta_k} \tag{4.29}$$

where $ENS^{\vartheta_k} = \sum_{i \in N}\overline{LS}_i$ and $RP^{\vartheta_k} = \sum_{i \in N}RP_i$.

ENS and RP are computed for the three cases above. The expected ENS (EENS) and expected RP (ERP) are estimated from the multiple MCS-OPF realizations. The steps for computing *EENS*, *ERP* and *EACG* using the MCS-OPF framework are sketched in Figure 4.11.

4.5 OPF of DGSs Against Unreliable Network

4.5.1 Settings of Networked DGSs

The proposed simulation and assessment framework are demonstrated on an integrated system of systems adapted from the *IEEE 13 node test feeder*, sketched in Figure 4.12 and equipped with the communication networks defined by (4.1) to (4.8). The conventional Grey model GM (1, 1) is chosen as the prediction tool, to support operation in case of degraded communication networks.

The DGS keeps the original radial structure but neglects the feeder of length zero, the regulator, the capacitor and the switch [349]. The final network has 11 nodes and the relevant characteristics of interest for the analysis, such as the presence of a main power supply spot, comparatively low and high spot and distributed loads [350]. The node index is denoted as i, MS indicates the main supply spot and $\downarrow$ indicates the spot load (kW).

Table 4.1 Configurations of the communication networks types.

CNs*	Transmission Delay (s)			Packet dropout
	LB*	MADT*	UB*	
N1	3	13	14	[? ? 0; 0.9 0 0.1; 1 0 0]
N2	3	13	14	[? ? 0; 0.7 0 0.3; 1 0 0]
N3	2	12	13	[0.95 0.05 0 0; ? 0 ? 0; 0.85 0 0 0.15; 1 0 0 0]
N4	2	12	13	[0.9 0.1 0 0; ? 0 ? 0; 0.75 0 0 0.25; 1 0 0 0]
N5	2	12	13	[0.85 0.15 0 0; ? 0 ? 0; 0.7 0 0 0.3; 1 0 0 0]

where CN: communication networks, LB: lower bound, MADT: maximum allowable transmission delay, UB: upper bound

The communication networks provide the data exchange between each power source and the CCC. Since real communication networks are usually mixtures of LANs, HANs and WANs, they have various configurations under different performance indices.

Table 4.1 gives the configurations of the different types of communication networks with respect to lower/upper bound, maximal allowable delay time and uncertain transition probability matrix [311, 317, 318]. Two simulation results are shown in Figure 4.2 and Figure 4.3. The power coefficient a used in (4.2) is set to 0.2041, based on reference [311].

As mentioned, the structure of DGSs is radial with 11 nodes and 10 feeders, as shown in Figure 4.12. The nominal voltage V is 4.16 kV, remaining constant for the resolution of the DC-OPF problem. Table 4.2 gives the technical characteristics of each feeder, consisting of the pairs of nodes (i, i') it connects, its length l, reactance X^{FD}, ampacity A^{FD}, failure rate λ^F, repair rate λ^R and variable operation cost Cov [329, 349].

The simulation study is performed on GRID DATA, whose horizon is from January 1, 2015 to September 22, 2015. The nodal power demands are directly obtained from the load data of this database. Necessary scale down of the GRID DATA to the level of *IEEE 13 node test feeders* is conducted. The daily load profiles of different nodes are shown in Figure 4.13.

Figure 4.12 Radial 11-nodes DGS.

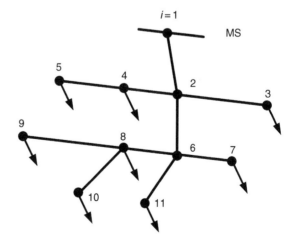

Table 4.2 Technical characteristics different feeders.

Type	Node i	Node i'	l(km)	X^{FD} (Ω/km)	A^{FD} (A)	
T1	1	2	0.610	0.371	730	
T2	2	3	0.152	0.472	340	
T3	2	4	0.152	0.555	230	
T1	2	6	0.610	0.371	730	
T3	4	5	0.091	0.555	230	
T6	6	7	0.152	0.252	329	
T4	6	8	0.091	0.555	230	
T1	6	11	0.305	0.371	730	
T5	8	9	0.091	0.555	230	
T7	8	10	0.244	0.318	175	

Type	Node i	Node i'	λ^F(1/h)	λ^R(1/h)	Cov($)	CopD($)
T1	1	2	3.333e-04	0.198	1.970e-02	0.02
T2	2	3	4.050e-04	0.162	9.173e-03	0.02
T3	2	4	3.552e-04	0.185	6.205e-03	0.02
T1	2	6	3.333e-04	0.198	6.205e-03	0.02
T3	4	5	3.552e-04	0.185	6.205e-03	0.02
T6	6	7	4.048e-04	0.164	8.904e-03	0.02
T4	6	8	3.552e-04	0.185	1.970e-02	0.02
T1	6	11	3.333e-04	0.198	1.970e-02	0.02
T5	8	9	3.552e-04	0.185	9.173e-03	0.02
T7	8	10	3.552e-04	0.185	6.205e-03	0.02

The power profiles of PV, WTG and PHEV demonstrated by the examples in Figure 4.8 and Figure 4.9, are also adapted from GRID DATA. They are used as the inputs to the MCS-OPF model. The values of the technical parameters, the failure rates, repair rates, costs of the MS and four types of DG technologies (PV, WTG, PHEV and ST), rated power of PHEV and ST, and charging level of ST, are taken from [329, 351] and reported in Table 4.3.

The maximum value of the energy price ep^{ϑ_k} is set to 0.12 ($/kWh) and the highest value of total demand OL_h is set to 4800 (kW), following [329]. The penalty cost for kWh of energy not supplied Cop is set equal to twice the maximum energy price. The penalty cost for kWh of redundant power $CopR$ is related to the real-time energy price and equals to $\frac{3}{4}ep^{\vartheta_k}$. The re-distributing cost for the total kWh handled by the automatic protection mechanism $CopD$ is related to the operational cost of each feeder and set to 0.02 ($/kWh).

Note that at the DGS level, the characteristics of solar irradiation, wind speed and behavior of PHEV are assumed uniform across the whole system, i.e., the power profiles of PV, WTG and PHEV at different nodes are almost identical, with only small variations.

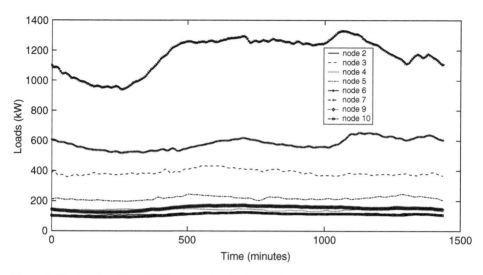

Figure 4.13 Load profiles of different nodes for August 23, 2015.

Table 4.3 Technical characteristics of the different power sources.

Type	Technical parameters	Costs
MS	$\lambda^F = 4.00\text{e} - 04(1/\text{h})$ $\lambda^R = 1.30\text{e} - 02(1/\text{h})$	$Cov = 0.145(\$/\text{kWh})$
PV	$\lambda^F = 5.00\text{e} - 04(1/\text{h})$ $\lambda^R = 1.30\text{e} - 02(1/\text{h})$	$C_i = 48(\$)$ $Cov = 3.76\text{e} - 05(\$/\text{kWh})$
W	$\lambda^F = 6.00\text{e} - 04(1/\text{h})$ $\lambda^R = 1.30\text{e} - 02(1/\text{h})$	$C_i = 113{,}750(\$)$ $Cov = 0.039(\$/\text{kWh})$
PHEV	$CL^{PHEV} = 1.4(\text{kW})$ $\lambda^F = 2.00\text{e} - 04(1/\text{h})$ $\lambda^R = 9.70\text{e} - 02(1/\text{h})$	$C_i = 17{,}000(\$)$ $Cov = 0.022(\$/\text{kWh})$
ST	$P_R^{ST} = 10(\text{kW})$ $Q_n^{ST} = 20(\text{kWh})$ $\lambda^F = 3.00\text{e} - 04(1/\text{h})$ $\lambda^R = 7.30\text{e} - 02(1/\text{h})$	$C_i = 135.15(\$)$ $Cov = 4.62\text{e} - 05(\$/\text{kWh})$

Based on (4.9), one random realization of the configuration of different power sources deployed in the DGS is given by the following matrix:

$$
\Xi =
\begin{pmatrix}
MS & PV & W & EV1 & EV2 & EV3 & EV4 & EV5 & EV6 & ST \\
1 & 0 & 0 & 0 & 0 & 0 & 0 & 0 & 0 & 0 \\
0 & 1 & 2 & 0 & 0 & 0 & 0 & 0 & 0 & 1 \\
0 & 0 & 1 & 0 & 0 & 0 & 0 & 0 & 0 & 1 \\
0 & 0 & 0 & 0 & 0 & 3 & 0 & 0 & 0 & 1 \\
0 & 1 & 0 & 0 & 0 & 0 & 0 & 0 & 0 & 1 \\
0 & 1 & 1 & 0 & 0 & 0 & 0 & 0 & 0 & 1 \\
0 & 2 & 0 & 1 & 0 & 0 & 0 & 0 & 0 & 0 \\
0 & 1 & 0 & 0 & 0 & 0 & 0 & 1 & 0 & 1 \\
0 & 1 & 1 & 0 & 1 & 0 & 0 & 0 & 0 & 0 \\
0 & 0 & 1 & 0 & 0 & 0 & 0 & 0 & 0 & 1 \\
0 & 1 & 0 & 0 & 0 & 0 & 2 & 0 & 0 & 1
\end{pmatrix}
\tag{4.30}
$$

where $EV1 \cdots EV6$ stand for the vehicles of PHEV20 Time Scaling, PHEV30 Power Scaling, PHEV30 Time Scaling, PHEV30 Power Scaling, PHEV40 Time Scaling and PHEV40 Power Scaling, respectively. Their daily charging load profiles have been illustrated in Figure 4.9.

4.5.2 OPF Under Different Demand Levels

The first study on the cases previously indicated is conducted representing different situations of power demand level: peak level (11:00-14:00 and 18:00-24:00), slack level (24:00-7:00) and flat level (7:00-11:00 and 14:00-18:00).

The communication networks configuration N1 is adopted in this illustrated study. The duration of each MCS can last for one or more hours. In this chapter, it is set as 1 hour since the data exchange rate is quite high and the hourly global cost of the integrated system is appropriate for practical applications.

For each power demand level, one duration is randomly sampled. The simulation results of each duration are shown in Table 4.4. The performance of data prediction is acceptable considering the difference between the various indicators of case 1 and case 2. As expected, the real-time operational cost increases significantly considering packet dropout in the communication networks [129, 139, 141]. The main reason is that the data packet forces CCC to use inadequate information for power dispatch.

Additionally, the packet dropout renders the local power sources unable to receive the command signal, and, so, they use inaccurate information to determine their power outputs. As a consequence, underestimated/overestimated power demand and excessive/insufficient power production lead to significant redundant power or load shedding.

The results of AGC^{θ_k} ($), ENS^{θ_k}(kW) and RP^{θ_k}(kW) for each scenario of the flat period are given in Table 4.5. Also, the value of GC^{θ_k}($) obtained from the OPF model at CCC is provided for comparison.

Table 4.4 Simulation results of different cases under different demand levels.

Demand Level	Case	AGC^ϑ ($)	ENS^ϑ (kW)	RP^ϑ (kW)
Peak	1	1000.8071	174.7792	0
	2	1073.4818	2293.0222	0
	3	1295.7314	2481.2697	0
Flat	1	483.2443	33.9741	0
	2	500.9015	441.6357	0
	3	531.4515	430.9372	29.1733
Slack	1	94.7867	2.5557	0
	2	349.2322	706.2998	0
	3	434.9720	766.3103	0

It is seen that $AGC2^{\vartheta_k}$ is always larger than $AGC1^{\vartheta_k}$. This is due to the fact that the data prediction made by CCC is not always accurate, thus, possibly overestimating/underestimating power loads and available power for each power source. For the scenario where $AGC2^{\vartheta_k}$ equals to $AGC3^{\vartheta_k}$, it means that there are no packet dropouts occurring in all data communication from CCC to local power sources.

$AGC3^{\vartheta_k}$ is always larger than $AGC2^{\vartheta_k}$, because the local power sources are unable to predict the dispatching command from CCC and have to use false command due to packet dropouts. A larger difference between $AGC3^{\vartheta_k}$ and $AGC2^{\vartheta_k}$ indicates that the packet dropout occurs with higher frequency in the corresponding scenario.

4.5.3 OPF Under Entire Period

Furthermore, $NS = 500$ random samples, each containing 12 consecutive scenarios with OPF, are generated over GRID DATA. Results of EAGC, EENS and ERP are reported in Table 4.6 for different types of communication networks; the highlighted values are used for comparison studies in the following paragraphs.

For case 1 with perfect communication networks, the differences among various estimated $EAGC1$ or $EENS1$ are relatively small (below $\pm5\%$) and come from the uncertainties of the renewable DGs and the contingencies.

For case 2 with data prediction, the results are also similar. The difference between $EAGC1$ and $EAGC2$ is introduced by the data prediction for next period due to imperfect communication networks. In order to minimize the difference, there is a need to further develop the performance of the prediction method, by future research work.

For case 3 without data prediction, one important observation is retained. The value of $EAGC3$ increases considerably with more packet dropouts in the communication networks.

Table 4.5 CG, Actual Global Cost (AGC), ENS and RP for each scenario of the peak period.

Case	1		2			
Type	$AGC1^{\vartheta_k}$	$ENS1^{\vartheta_k}$	$GC2^{\vartheta_k}$	$AGC2^{\vartheta_k}$	$ENS2^{\vartheta_k}$	$RP2^{\vartheta_k}$
1	47.6694	5.3595	46.8184	49.2365	66.8314	0
2	47.3965	5.2532	46.8184	49.1472	65.7501	0
3	47.7261	5.3606	45.7441	49.4142	67.0153	0
4	47.6138	5.3008	46.6442	48.8337	65.1072	0
5	46.381	4.8508	47.7097	47.0554	58.1258	0
6	43.2764	3.7648	47.8205	43.6794	44.2721	0
7	39.4250	2.4275	45.9061	39.8867	28.6351	0
8	35.8970	1.2166	41.4582	36.6345	15.4274	0
9	33.5928	0.4403	36.1528	34.6978	7.8178	0
10	31.3838	0	31.6467	33.2922	5.0574	0
11	31.3704	0	31.996	35.0810	10.8118	0
12	31.5121	0	31.8516	33.9429	6.7843	0
Sum	483.2443	33.9741	500.5667	500.9015	441.6357	0

Case	3			
Type	$GC3^{\vartheta_k}$	$AGC3^{\vartheta_k}$	$ENS3^{\vartheta_k}$	$RP3^{\vartheta_k}$
1	46.8184	49.2365	66.8314	0
2	46.8184	49.1472	65.7501	0
3	46.6879	49.4164	67.0153	0
4	46.3843	48.8834	65.1072	0
5	46.947	54.9172	61.2410	0
6	46.6791	46.2771	44.7509	0
7	45.4748	41.8279	27.7879	0
8	42.1276	43.0927	12.9591	0
9	39.3837	44.5816	7.8178	0
10	32.3798	34.3382	0	15.6439
11	31.6626	32.9331	0	13.5294
12	32.1164	36.8002	11.6765	0
Sum	503.48	531.4515	430.9372	29.1733

N2 has higher probability to have more packet dropouts than N1. Therefore, the value of *EAGC*3 of N2 is much larger than that of N1.

The comparisons among the values of *EAGC*3 of N3, N4 and N5 confirm this finding. The order of degradation levels of the three communication networks is N5 > N4 > N3, with respect to the packet dropouts.

Table 4.6 EACG, EENS and ERP for the entire duration of the database.

CNs Indicator	N1	N2	N3	N4	N5
EAGC1	179.7009	177.1943	179.5252	181.9325	174.8535
EENS1	106.9149	104.1753	107.7122	112.7846	104.3514
EAGC2	291.8887	291.4971	304.4781	315.3020	314.9378
EENS2	192.6631	196.5827	188.8514	198.0111	188.8513
ERP2	29.8	30.2	30.6	29.9	29.8
EAGC3	352.5611	385.9817	340.2547	367.9292	390.6081
EENS3	201.2433	204.2968	199.4828	209.1599	201.4725
ERP3	34.4188	38.3266	36.3	36.4228	35.3

Indeed, from the perspective of the CC already the measurements unavailability at the nodes or the packet dropouts would imply receiving incomplete data set. Consequently, the outcome of the analysis on the impact caused by packet dropouts corresponds to measurement unavailability.

The results for case 3 with imperfect communication networks in Table 4.6 demonstrate that the expected actual global cost is much higher in the situation with more packet dropouts, which implies that the unavailability of measurements at each node could have similar effects.

To handle this, we have introduced a grey differential model-based prediction method to estimate the missing measurements for obtaining a complete data set. In Table 4.6, the comparison between case 2 with prediction and case 3 without prediction proves that the proposed data prediction method works well for the missing measurement due to packet dropout. Therefore, we believe that the prediction method can also be used to mitigate the influences caused by different levels of measurements availability.

The simulation is run on a 64-bit Windows desktop with 32 GB memory and an Intel(R) Xeon(R) E5-1650 v3 @ 3.50GHz CPU. In each scenario, the OPF model is described by a linear equation system which can be solved by the Simplex method, which is a linear programming (LP) algorithm with polynomial-time complexity and takes around 1 second.

The running time for obtaining all indicators of *IEEE 13 node test feeder* in 12 consecutive scenarios is 15 seconds. To obtain the expectation of each indicator, the MCS-OPF contains 500 samples and can be finished in 2 hours. As the MCS-OPF method is offline, the running time is not a major concern. Since the time complexity grows polynomially along with the LP problem size, this approach is still promising in the case of more practical constraints and more variables.

5

Maintenance of Aging CPSs

Aging process is inevitable for any component in Cyber-Physical Systems (CPSs). The output of control systems, e.g., the power system frequency in this chapter, is much more readily measurable as compared to the state of the components [18, 160–163, 172, 352]. Moreover, the failure condition is directly determined by the frequency deviation resulting from the Load Frequency Control (LFC) performance [18, 160, 162, 164]. Therefore, this chapter extends the Condition-Based Maintenance (CBM) model used in [167, 173, 353–355], by considering the performance-based maintenance (PBM) model in which maintenance activities only depend on the reduced LFC performance.

The degraded control system receives a corrective maintenance action if the current control performance exceeds the failure thresholds; on the other hand, it receives a preventive maintenance action if the current control performance exceeds the preventive thresholds, in order to avert the occurrence of failures when the system can still operate [353–355]. The PBM is then employed in the degraded power system with four gas turbine units and LFC tuned for rejection ability [18].

Maintenance strategies are compared by estimating the long-term maintenance cost rate (MCR), for different preventive maintenance thresholds and inspection intervals [165, 355] using Monte Carlo simulation (Monte Carlo simulation (MCS) [356]. Finally, the optimum maintenance strategy minimizing MCR is identified, which combines maintenance and safety (downtime) costs. As the MCR lacks an explicit formulation due to the feedback control mechanisms [171, 176], the closed-form solution of the optimal maintenance strategy is unavailable.

5.1 Data-driven Degradation Model for CPSs

This section introduces the degradation model of the control system, presents the degradation path model with random starting time and unit-to-unit variability, and, finally, discusses the failure criteria employed in the power system.

5.1.1 Degraded Control System

The developed degradation model for control systems is general, and in this section, it is exemplified with reference to electric power systems. This proposed degradation model

Part of the contents are from [294, 366] and permission has been obtained to use the contents.

Cyber-Physical Distributed Systems: Modeling, Reliability Analysis and Applications, First Edition.
Huadong Mo, Giovanni Sansavini and Min Xie.
© 2021 John Wiley & Sons Ltd. Published 2021 by John Wiley & Sons Ltd.

works for linear and nonlinear control systems, and for other types of industries, e.g., manufacturing systems [152] and chemical plants [171].

The investigated power system subject to aging process is a natural gas power plant (Lake Cogen Ltd, 60.5 MW) [357], which consists of four Rolls-Royce industrial AVON Mk 1535 gas turbines (15 MW). In this system, generation and load are balanced in real-time upon sudden load increases by enforcing the LFC strategy presented in Figure 5.1. However, the degradation of compressors and sensors reduces the LFC performance and can lead to failures in the power system.

The detailed representation is provided for Gas Turbine 2 and is representative of the other units as well. The dynamics of the governor, compressor combustor and turbine [160] are given in transfer functions as

$$\text{Governor: } G_g = \frac{1}{cs + b} \tag{5.1}$$

$$\text{Compressor: } G_{cp} = \frac{Xs + 1}{Ys + 1} \tag{5.2}$$

$$\text{Combustor: } G_{cb} = \frac{T_c s + 1}{T_f s + 1} \tag{5.3}$$

$$\text{Turbine: } G_t = \frac{1}{T_t s + 1} \tag{5.4}$$

where b and c are the gas turbine constant of the fuel valve positioner. X and Y are the lead time constant and the lag time constant of the gas turbine speed governor. T_c and T_f are the combustion reaction time delay and the gas turbine fuel time constant. T_t is the compressor discharge time constant.

The frequency deviation of the power system Δf depends on the difference between power generated and the sudden load increase, i.e., $\sum_{n=1}^{4} \Delta P_G^n - \Delta P_d$

$$\Delta f = \frac{K_p}{T_p s + 1} \cdot \left(\sum_{n=1}^{N} \Delta P_G^n - \Delta P_d \right) \tag{5.5}$$

where ΔP_G^n is the output of gas turbine n and ΔP_d is the sudden load increase. K_p and T_p are the gain and the time constant of the power system. B and R are the frequency bias and the governor speed regulation parameter, respectively.

The Phasor Measurement Units (PMU) measures the system frequency and sends it to the controller every interval T_s. Therefore, realistic power systems employ digital controllers for implementing the LFC strategy. The digital controller $C(z)$ is expressed in the discrete-time formulation as

$$C(z) = K_P \left(1 + \frac{F(z)}{K_I} + K_D \cdot \frac{N}{1 + NF(z)} \right) \tag{5.6}$$

where K_P, K_I and K_D are the proportional gain, integral gain and derivative gain of the Proportional–Integral–Derivative (PID) controller. Our work adopts the forward Euler method and therefore $F(z) = \frac{T_s}{z-1}$.

The degradation models for aging actuator and sensor have been illustrated in Section 2.1.1.2. Therefore, the developed control-block-diagram-based model with aging components can predict the system performance at any operating time.

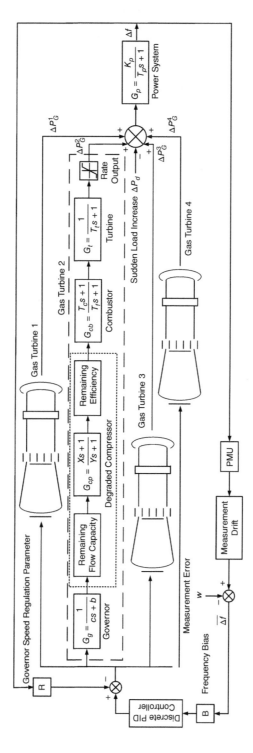

Figure 5.1 Control block diagram of the power system with four gas turbines subject to aging process.

5.1.2 Parameter Estimation via EM Algorithm

In this section, the Expectation-Maximization (EM) algorithm employed in the determination of the degradation parameters from the available data is described.

Based on the EM algorithm, the posterior estimator of λ_n in the degradation model is updated once new observations are available.

- **Step 1**: In (2.3), the stochastic parameter λ_n follows a prior distribution $p(\lambda_n)$, which is normally distributed with mean u_λ and variance σ_λ^2 [172]. Based on the properties of the standard Brownian motion, for a given λ_n, then sampling distribution of $\boldsymbol{Y}_{n:k} = \{y_{n,1}, y_{n,2}, \cdots, y_{n,k}\}$ is multi-variate normal

$$p(\boldsymbol{Y}_{n:k}|\lambda_n) = \frac{1}{\prod_{j=1}^{k} \sqrt{2\pi\sigma_B^2(\overline{T_{n,j}} - \overline{T_{n,j-1}})}} \exp\left(-\sum_{j=1}^{k} \frac{(y_{n,j} - y_{n,j-1} - \lambda_n(\overline{T_{n,j}} - \overline{T_{n,j-1}})^2)}{2\sigma_B^2(\overline{T_{n,j}} - \overline{T_{n,j-1}})}\right)$$

(5.7)

To compute the posterior $p(\lambda_n|\boldsymbol{Y}_{n:k})$ in the Bayesian framework, it is assumed that the prior distribution of λ_n is $N(u_\lambda, \sigma_\lambda^2)$, which belongs to the conjugate family of the sampling distribution $p(\boldsymbol{Y}_{n:k}|\lambda_n)$ [179]. As a result, the posterior estimator of λ_n conditional on $\boldsymbol{Y}_{n:k}$ is also normal, i.e., $\lambda_n | \boldsymbol{Y}_{n:k} \sim N(u_{\lambda_n,k}, \sigma_{\lambda_n,k}^2)$. Other prior distributions are also applicable, however the evolution of the posterior may require numerical techniques such as Gibbs sampling [180].

- **Step 2**: Once a new observation at $\overline{T_{n,k}}$ is available, the posterior distribution of λ_n is updated based on Bayes theorem

$$p(\lambda_n|\boldsymbol{Y}_{n:k}) = \frac{p(\boldsymbol{Y}_{n:k}|\lambda_n) \cdot p(\lambda_n)}{p(\boldsymbol{Y}_{n:k})} \propto p(\boldsymbol{Y}_{n:k}|\lambda_n) \cdot p(\lambda_n).$$

(5.8)

Given $\lambda_n \sim N(u_\lambda, \sigma_\lambda^2)$ and Eq. (5.8), the posterior $p(\lambda_n|\boldsymbol{Y}_{n:k})$ is computed as

$$p(\lambda_n|\boldsymbol{Y}_{n:k}) \propto p(\boldsymbol{Y}_{n:k}|\lambda_n) \cdot p(\lambda_n)$$

$$\propto \exp\left(-\sum_{j=1}^{k} \frac{(y_{n,j} - y_{n,j-1} - \lambda_n(\overline{T_{n,j}} - \overline{T_{n,j-1}})^2)}{2\sigma_B^2(\overline{T_{n,j}} - \overline{T_{n,j-1}})}\right) \cdot \exp\left(-\frac{(\lambda_n - u_\lambda)^2}{2\sigma_\lambda^2}\right)$$

$$\propto \exp\left(-\frac{(\lambda_n - u_{\lambda,k})^2}{2\sigma_{\lambda,k}^2}\right)$$

(5.9)

and because of the properties of the normal distribution $p(\lambda_n|\boldsymbol{Y}_{n:k})$

$$p(\lambda_n|\boldsymbol{Y}_{n:k}) = \frac{1}{\sqrt{2\pi\sigma_{\lambda,k}^2}} \exp\left(-\frac{(\lambda_n - u_{\lambda,k})^2}{2\sigma_{\lambda,k}^2}\right)$$

(5.10)

with

$$u_{\lambda_n,k} = (u_\lambda\sigma_B^2 + y_{n,k}\sigma_\lambda^2)/(\overline{T_{n,k}}\sigma_\lambda^2 + \sigma_B^2)$$

(5.11)

$$\sigma_{\lambda_n,k}^2 = \sigma_B^2\sigma_\lambda^2/(\overline{T_{n,k}}\sigma_\lambda^2 + \sigma_B^2)$$

(5.12)

- **Step 3**: in this step, $\Theta_{n,k} = [\sigma^2_{B,k}, u_{\lambda,k}, \sigma^2_{\lambda,k}]$ is estimated. Based on the EM algorithm [179], the completed log-likelihood function of $p(Y_{n:k}, \lambda_n | \Theta_{n,k})$ is first determined

$$\ln p(Y_{n:k}, \lambda_n | \Theta_{n,k}) = \ln p(Y_{n:k}|\lambda_n, \Theta_{n,k}) + \ln p(\lambda_n|\Theta_{n,k}) = -\frac{k+1}{2}\ln 2\pi$$

$$-\frac{1}{2}\sum_{j=1}^{k} \ln(\overline{T_{n,j}} - \overline{T_{n,j-1}}) - \frac{k}{2}\ln\sigma^2_{B,k} - \sum_{j=1}^{k}\frac{(y_{n,j} - y_{n,j-1} - \lambda_n(\overline{T_{n,j}} - \overline{T_{n,j-1}})^2)}{2\sigma^2_{B,k}(\overline{T_{n,j}} - \overline{T_{n,j-1}})} \quad (5.13)$$

$$-\frac{1}{2}\ln\sigma^2_{\lambda,k} - \frac{(\lambda_n - u_{\lambda,k})^2}{2\sigma^2_{\lambda,k}}.$$

Given $\widehat{\Theta}^{(i)}_{n,k} = [\widehat{\sigma}^{2(i)}_{B,k}, \widehat{u}^{2(i)}_{\lambda,k}, \widehat{\sigma}^{2(i)}_{\lambda,k}]$, i.e., the estimator obtained in the i-th step based on $Y_{n:k}$, the expectation $\ell\left(\Theta_{n,k}|\widehat{\Theta}^{(i)}_{n,k}\right)$ of $\ln p(Y_{n:k}, \lambda_n | \Theta_{n,k})$ is calculated

$$\ell\left(\Theta_{n,k}|\widehat{\Theta}^{(i)}_{n,k}\right) = E_{\lambda_n|Y_{n:k}, \widehat{\Theta}^{(i)}_{n,k}}\{\ln p(Y_{n:k}, \lambda_n|\Theta_{n,k})\} = -\frac{k+1}{2}\ln 2\pi \quad (5.14)$$

$$-\frac{1}{2}\sum_{j=1}^{k}\ln(\overline{T_{n,j}} - \overline{T_{n,j-1}}) - \frac{k}{2}\ln\sigma^2_{B,k} - \frac{1}{2}\ln\sigma^2_{\lambda,k} - \frac{u^2_{\lambda_n,k} + \sigma^2_{\lambda_n,k} - 2u_{\lambda_n,k}u_{\lambda,k} + u^2_{\lambda,k}}{2\sigma^2_{\lambda,k}}$$

$$-\sum_{j=1}^{k}\frac{(y_{n,j} - y_{n,j-1})^2 - 2u_{\lambda_n,k}(y_{n,j} - y_{n,j-1})(\overline{T_{n,j}} - \overline{T_{n,j-1}}) + (u^2_{\lambda_n,k} + \sigma^2_{\lambda_n,k})(\overline{T_{n,j}} - \overline{T_{n,j-1}})^2)}{2\sigma^2_{B,k}(\overline{T_{n,j}} - \overline{T_{n,j-1}})}$$

Let $\frac{\partial\ell\left(\Theta_{n,k}|\widehat{\Theta}^{(i)}_{n,k}\right)}{\partial\Theta_{n,k}} = 0$, $\widehat{\Theta}^{(i+1)}_{n,k}$ can be computed as

$$\widehat{\sigma}^{2(i+1)}_{B,k} =$$

$$\frac{1}{k}\sum_{j=1}^{k}\frac{(y_{n,j} - y_{n,j-1})^2 - 2u_{\lambda_n,k}(y_{n,j} - y_{n,j-1})(\overline{T_{n,j}} - \overline{T_{n,j-1}}) + (u^2_{\lambda_n,k} + \sigma^2_{\lambda_n,k})(\overline{T_{n,j}} - \overline{T_{n,j-1}})^2)}{(\overline{T_{n,j}} - \overline{T_{n,j-1}})}$$

$$\qquad\qquad\qquad\qquad\qquad\qquad\qquad\qquad\qquad\qquad\qquad\qquad (5.15)$$

$$\widehat{u}^{(i+1)}_{\lambda,k} = u_{\lambda_n,k} \qquad\qquad\qquad\qquad\qquad\qquad\qquad\qquad (5.16)$$

$$\widehat{\sigma}^{2(i+1)}_{\lambda,k} = \sigma^2_{\lambda_n,k}. \qquad\qquad\qquad\qquad\qquad\qquad\qquad (5.17)$$

5.1.3 LFC Performance Criteria

The performance of the degrading control system is successful, if it satisfies the three general criteria, i.e., Rising Time (RT), Percentage Overshoot (PO) and ST in Section 2.2.1.1 [171, 175, 176]; otherwise, the control system fails.

In electric power systems, a standard procedure is followed to verify whether the system adopting the LFC strategy operates satisfactorily against a sudden load increase in the face of aging. Based on [164], if one of the two features related to the system frequency exceeds the maximum allowable value, the LFC strategy fails to stabilize the system frequency and the power system is down. These features are:

- maximum transient frequency drop (undershoot frequency) $f_{d,\,max}$ (mHz) due to the specified sudden load increase, which is quantified by PO;

- frequency recovery time after load increase $f_{f,in}$ (s), the time interval between the departure from the steady-state frequency band after the specified sudden load increase and the permanent re-entry of the frequency into the steady-state frequency tolerance band. This feature is quantified by ST.

UCTE policies prescribe that the sudden load increase ΔP_d is 0.01 pu, the maximum allowed frequency drop is -800 mHz, the steady-state frequency tolerance band is ± 20 mHz and the maximum allowed recovery time is 30 s [308]. In this chapter, the base value for the apparent power is 3000 MVA and the rated frequency is 50 Hz.

The values of PO and ST are both predicted from the power system frequency curve, which is the output of the simulation model presented in Section 5.2.1.

5.2 Maintenance Model and Cost Model

5.2.1 PBM Model

In the proposed PBM model, maintenance activities are scheduled following the reduction of control performance. The degraded control system is inspected every time interval T_I to assess performance deterioration.

In this section, at inspection time kT_I ($k = 1, 2, \ldots$), the preventive replacement is carried out on the power system if at least one of two aforementioned features exceeds the preventive maintenance thresholds M_{PO} and M_{ST}, respectively. On the other hand, the corrective replacement is performed if the LFC fails to satisfy the maximum allowed frequency drop or recovery time, i.e., if at least one of the failure thresholds L_{PO} and L_{ST} is exceeded, respectively.

Corrective and preventive maintenance actions entail a major overhaul of the aging compressor (i.e., recoating of the blades, stator rings and vanes, repair of the rotor seals, and repair of the compressor rotor etc.) and the calibration of the PMU. It is assumed that the system is restored to its original conditions after maintenance. Therefore, the sequence of maintenance actions defines a renewal process.

Similar to [167, 173, 353], the proposed PBM model consists of three variables:

- CM_k indicates whether the power system receives a corrective maintenance at time kT_I.
- PM_k indicates whether the power system receives a preventive maintenance at time kT_I.
- Dt_k is the system downtime before time kT_I, i.e., the amount of time between two consecutive inspections during which the power system cannot ensure satisfactory LFC performance.

These maintenance variables can be expressed as

$$X_{M,L,T_I} \triangleq \{X_{M,L,T_I}(k) \triangleq (CM_k, PM_k, Dt_k)\} \tag{5.18}$$

where $M = [M_{PO}, M_{ST}]$ and $L = [L_{PO}, L_{ST}]$ are the preventive maintenance thresholds and the failure thresholds, respectively.

The variables of the PBM model evolve at each inspection k occurring at time kT_I, and their values are updated as

$$PM_k = \begin{cases} 1 \text{ if } (k-1)T_I < T_p^k \leq kT_I \text{ and } kT_I < T_f^k \\ 0 \qquad\qquad\qquad\qquad otherwise \end{cases} \tag{5.19}$$

$$CM_k = \begin{cases} 1 \ if \ (k-1)T_I < T_p^k \ and \ (k-1)T_I < T_f^k \le kT_I \\ 0 \qquad\qquad\qquad\qquad\qquad\qquad otherwise \end{cases} \tag{5.20}$$

$$Dt_k = (kT_I - T_f^k)CM_k \tag{5.21}$$

where T_p^k and T_f^k are the time instants at which the control system exceeds the preventive and the failure thresholds, respectively, between two consecutive $k-1$ and k inspections computed as

$$T_p^k = \min\{T_{p_{PO}}^k, T_{p_{ST}}^k\} \tag{5.22}$$

$$T_f^k = \min\{T_{f_{PO}}^k, T_{f_{ST}}^k\} \tag{5.23}$$

where $T_{p_{PO}}^k$ and $T_{p_{ST}}^k$ are the time instants at which the features *PO* and *ST* exceed the preventive thresholds between inspection $k-1$ and k; $T_{f_{PO}}^k$ and $T_{f_{ST}}^k$ are the time instants at which the features *PO* and *ST* exceed the failure thresholds between inspection $k-1$ and k.

Figure 5.2 illustrates the evolution of the maintenance variables of the PBM model, i.e., PM_k, CM_k and Dt_k, as a function of the performance features, i.e., *PO* and *ST*, which determine T_p^k and T_f^k.

In this example, at inspection time T_I, *ST* reaches the preventive maintenance threshold, and T_p^1 is equal to $T_{p_{ST}}^1$. No feature reaches the failure threshold, and $T_f^1 > T_I$. Following (5.19), $PM_1 = 1$ and the preventive maintenance action is carried out. Following (5.20) and (5.21), $CM_1 = Dt_1 = 0$.

At inspection time $2T_I$, the controller features do not exceed the preventive maintenance and failure thresholds, and no maintenance is required. As a result, the power system can still operate normally, but the performance features keep deteriorating, and *PO* reaches the failure threshold before the inspection time $3T_I$. Therefore, $T_f^3 = T_{f_{PO}}^3$.

Following (5.19) and (5.20), $CM_3 = 1$ and a corrective maintenance is carried out at $3T_I$; following (5.21), $Dt_3 = 3T_I - T_f^3$. Figure 5.2 shows that $PM_k = 1$ and $CM_k = 1$ are mutually exclusive conditions, i.e., preventive maintenance cannot take place at inspection interval k, if corrective maintenance is performed.

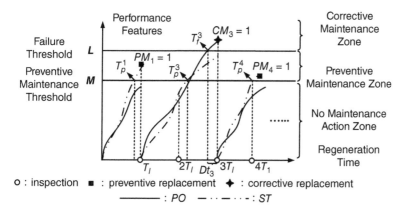

Figure 5.2 The evolution of the maintenance variables of the PBM model, i.e., PM_k, CM_k and Dt_k, as a function of the performance features, i.e., *PO* and *ST*.

5.2.2 Cost Model

The analysis of the maintenance cycle allows evaluating the cost efficiency of the PBM model because the overall maintenance costs depend on the number of maintenance actions which are carried out. The expected length of the maintenance cycle, i.e., the expected time before a preventive or a corrective maintenance action is carried out, is evaluated as

$$E[R_L] = \lim_{m \to \infty} \frac{mT_I}{\sum_{k=1}^{m}(PM_k + CM_k)} \tag{5.24}$$

where R_L is the maintenance cycle of the system, i.e., the time between two consecutive maintenance actions. The average cost per maintenance action is

$$C(\boldsymbol{M}, \boldsymbol{L}, T_I) = \lim_{m \to \infty} \frac{\sum_{k=1}^{m}(c_{PM}PM_k + c_{CM}CM_k + c_{DT}Dt_k + c_I)}{\sum_{k=1}^{m}(PM_k + CM_k)} \tag{5.25}$$

where c_{PM}, c_{CM}, c_{DT} and c_I are the unit costs, i.e., the cost factors, associated to preventive maintenance, corrective maintenance, system downtime and inspection activities, respectively.

Optimum maintenance strategies minimize the expected maintenance cost per unit time, i.e., the MCR, which is predicted by the long-term maintenance cost over the expected maintenance cycle length as

$$LC(\boldsymbol{M}, \boldsymbol{L}, T_I) = \frac{C(\boldsymbol{M}, \boldsymbol{L}, T_I)}{E[R_L]} \tag{5.26}$$

Due to the diverse stochastic degradation processes of each component and the feedback control loop structure, the explicit function for the MCR is unavailable. Therefore, MCS is employed to run the model for the degraded control system of Section 2.1.1.2, and to generate realistic scenarios of degraded gas turbines and PMU. The number of simulated inspection intervals is $m = 1500$, and $LC(\boldsymbol{M}, \boldsymbol{L}, T_I)$ is estimated using (5.25).

In addition, the availability of control systems quantifies the ability to deliver the expected control performance in the presence of degradation. The system availability A is defined as the ratio of the operation time and the inspection interval as

$$A = 1 - \frac{E[Dt]}{T_I} \tag{5.27}$$

$LC(\boldsymbol{M}, \boldsymbol{L}, T_I)$ is predicted via the MCS, therefore explicit expressions for the optimal preventive maintenance thresholds $\boldsymbol{M}$ and for the optimum inspection interval T_I are not available. In this situation, heuristics are widely used to optimize maintenance strategies, e.g., the maintenance of power system via Hybrid Genetic-Simulated-Annealing Algorithm (HGSAA) [264–266]. Therefore, HGSAA is adopted in this chapter, where the evolutionary algorithm and simulated annealing interact and exchange the best solutions in terms of $\boldsymbol{M}$ and T_I, achieving iteratively better maintenance strategies with reduced computation efforts as compared to similar optimization heuristics. The detailed procedure for the HGSAA implementation can be found in [264–266].

5.3 Applications to DGSs

The models for Distributed Generation Systems (DGSs) and different power sources, e.g., PV and WTG, have been explained in Section 4.3.1.

5.3.1 Output of Aging Generators

The validation study of the degradation models for aging generators, i.e., gas turbines, has been provided in the Section 2.2.2.

Given the scenario ϑ_k defined in Section 4.4.2, the Optimal Power Flow (OPF) model optimally dispatches the available power Pa^{ϑ_k} [kW] to supply the loads L_i^k [kW] in the DGS [124, 358]. At time t_k, the available power $Pa_{i,j}^k$ [kW] of power source type j at node i for scenario ϑ_k is:

$$Pa_{i,j}^k = \xi_{i,j}mc_{i,j}^k G_j(\vartheta_k) \tag{5.28}$$

where $\xi_{i,j}$ is the number of units of the power source type j deployed at node i; $G_j(\vartheta_k)$ is the output of the power source type j at scenario ϑ_k, computed in Section 4.3.1. For the MS type j at node i, $G_j(\vartheta_k) = P_{i,j}^{ms,k}$.

The degradation of generators efficiency and ESS reduces the maximum capacity Q^M, respectively. At time t_k, given the efficiency degradation level $X_{i,j,l}(t_k)$ of the unit l of power source type j at node i, the available power $Pa_{i,j,l}^k$ [kW] of unit l is:

$$Pa_{i,j,l}^k = mc_{i,j,l}^k G_j(\vartheta_k) \cdot (1 - X_{i,j,l}(t_k)) \tag{5.29}$$

where $mc_{i,j,l}^k$ is the mechanical state of the unit l of the power source type j at node i in scenario ϑ_k. Therefore, (5.28) is modified as:

$$Pa_{i,j}^k = \sum_{l=1}^{\xi_{i,j}} Pa_{i,j,l}^k \tag{5.30}$$

At time t_k, the evolution of Co_j^{PS} is determined by a Wiener process as shown in Section 2.1.1.2, i.e., the increment in Operation & Maintenance (O&M) cost obeys a Wiener process, as follows:

$$Co_j^{PS}(t_k) = Co_j^{PS}(t_k - \Delta t) + \overline{\lambda}b(t_k - \Delta t)^{b-1}\Delta t + \overline{\sigma}_B Y \sqrt{\Delta t} \tag{5.31}$$

5.3.2 Impact of Aging on DGSs

Since the energy price model and the fundamental OPF model have been provided in Section 4.4.3, the effects of aging process can be exemplified with reference to the IEEE 13 node test feeder.

5.3.2.1 Settings of Aging DGSs

The simplified aging DGS based on the IEEE 13 node test feeder [359] neglects regulators, capacitors, switches and the feeders of zero length in the linear DC approximation [89, 294, 322, 329]. As a result, the modified configuration given in Figure 4.12 encompasses 11 nodes, of which eight host loads, i.e., node 2, 3, 4, 5, 6, 7, 9 and 10.

The technical characteristics, i.e., length, reactance, ampacity, failure rate, repair rate and O&M costs, of each feeder (i, i') connecting nodes i and i' and energy price data are given in references [322, 346, 360]. The nominal voltage is 4.16 kV. Following [361], the modifications to the original IEEE 13 allow studying the integration of renewable Distributed Generation (DG) units.

Additionally, some locations and values of loads and ampacity values of the feeders have been modified to generate conditions of power congestion [322, 346, 360]. As a result, when there is no degradation, energy not supplied can occur at certain nodes due to the nature of renewable resources and the limitation of transmission lines.

Table 5.1 provides the name, the overall rated power, the number of units and the installation time of each generation plant at different nodes.

Therefore, the matrix in (4.9), which represents the configuration of DGSs in Corvallis, Oregon, U.S., is:

$$
\Xi = \begin{pmatrix}
G & B & W & PV & ESS \\
7 & 0 & 0 & 0 & 0 \\
0 & 3 & 0 & 5000 & 0 \\
0 & 0 & 0 & 8000 & 0 \\
0 & 0 & 0 & 8500 & 100 \\
0 & 0 & 7 & 0 & 50 \\
4 & 0 & 0 & 0 & 0 \\
0 & 0 & 8 & 0 & 0 \\
0 & 0 & 0 & 10000 & 150 \\
0 & 6 & 0 & 0 & 0 \\
0 & 0 & 5 & 0 & 100 \\
3 & 0 & 0 & 0 & 0
\end{pmatrix}.
\tag{5.32}
$$

Table 5.1 Technical details of generation plants.

Node	Power Plant	Overall Rated Power	Number of Units	Installation
			Characteristics	
1	Natural Gas Plant (Solano County)	2800 kW	7	2012.12
2	PV Farm (Yamhill Solar)	1000 kW	5000	2011.09
	Biomass Plant (POTB Digester)	900 kW	3	2013.10
3	PV Farm (Bellevue Solar)	1600 kW	8000	2011.09
4	PV Farm (Baldock Solar)	1700 kW	8500	2011.12
	ESS	200 kWh	100	2011.09
5	Wind Farm (Foundation AB)	1400 kW	7	2012.11
	ESS	100 kWh	50	2013.01
6	Natural Gas Plant (Oregon University)	1600 kW	4	2011.08
7	Wind Farm (Tooele)	1600 kW	8	2009.05
8	PV Farm (Black Solar)	2000 kW	10000	2011.12
	ESS	300 kWh	150	2012.02
9	Biomass Plant (Cargill)	1800 kW	6	2011.04
10	Wind Farm (Wal-Mart)	1000 kW	5	2012.09
	ESS	200 kWh	100	2012.11
11	Natural Gas Plant (Thomas M)	1200 kW	3	2011.12

Given the real-time weather data sampled from the dataset [362] with the same time stamp to keep dependence, the wind and photovoltaic power at one node is estimated.

One demonstrated collection of scenarios showing the renewable energy resources and load at different nodes is presented in Figure 5.3 for August 6, 2014, where the duration of each scenario is 15 minutes and the number of scenarios of this date is 96. Different nodes have different demand values as shown in Figure 5.3 (d). The SoC of ESS is given in Figure 5.3 (c), which demonstrates the variability of the charging pattern.

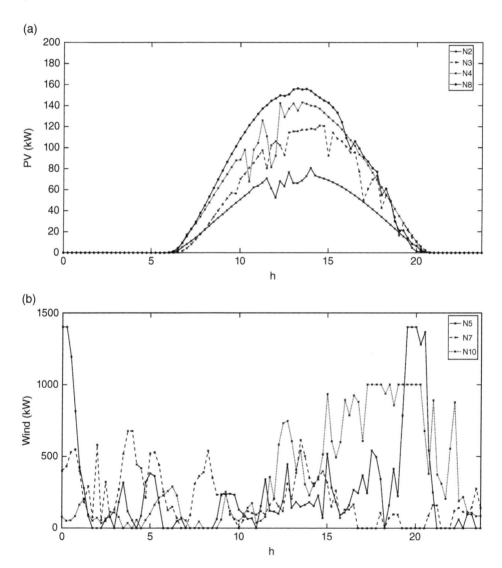

Figure 5.3 Scenario of the outputs of renewable energy resources, i.e., (a) PV farm and (b) wind farm, (c) SOC of ESS, and loads of (d) all nodes on August 6, 2014.

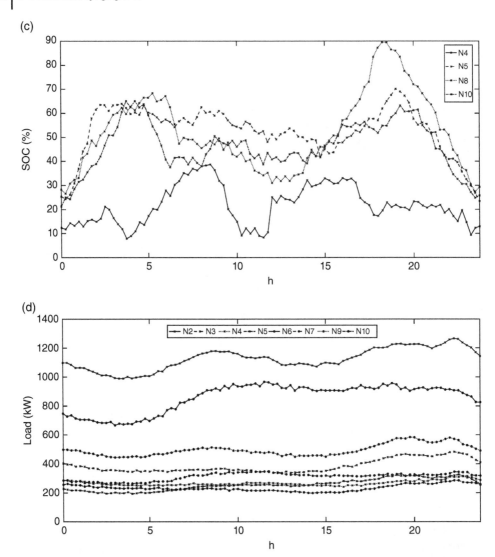

Figure 5.3 (*Continued*)

5.3.2.2 Validations of Generator Performance Indexes

Figure 5.4 shows the efficiency reduction (Figure 5.4 (a)), the O&M cost (excluding fuel) (Figure 5.4 (b)) and the overall O&M cost (Figure 5.4 (c)) for the natural gas plants Gadsden (historical data 1), operating since 1949, and Blount Station (historical data 2) [363], operating since 1968, for the linear, the quadratic and the Wiener degradation models.

Figure 5.4 (a) and Figure 5.4 (b) show that the efficiency drops and the O&M costs (excluding fuel) have similar increasing trends. In addition, the cross-correlation factor (CCF) between them for the gas plants Gadsden and Blount Station are 0.9891 and 0.9973, respectively. Additionally, Figure 5.4 (b) and Figure 5.4 (c) show that the O&M cost (excluding fuel) amounts at an increasing share of total O&M cost as the plant age increases, i.e., the percentage of the maintenance cost with respect to total O&M cost keeps increasing from

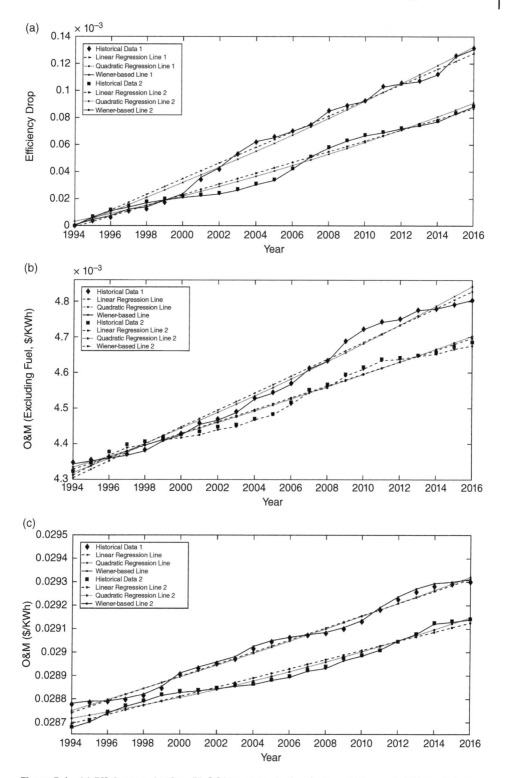

Figure 5.4 (a) Efficiency reduction, (b) O&M cost (excluding fuel) and (c) overall O&M cost for the natural gas plants Gadsden (historical data 1) and Blount Station (historical data 2) [355] computed via the linear regression model, the quadratic regression model and Wiener process-based model.

15.11% to 16.39% for the Gadsden station and from 15.08% to 16.08% for the Blount Station. This is due to the increasing number of unscheduled maintenance activities, caused by the decreasing efficiency.

The Wiener process-based model can better capture the evolution of the three performance indices, i.e., the efficiency, the O&M (excluding fuel) and the O&M cost, as compared to the linear and the quadratic regression lines. Additionally, the estimated Wiener-based lines in Figure 5.4 (a) to Figure 5.4 (c) are updated once a new observation is available. For example, if observations of O&M costs have been collected from year 1994 to 2010, then the O&M costs of 2011 can be predicted via the Wiener-based model based on these observations.

Table 5.2 provides the type of generator, rated power, efficiency reduction model and O&M cost model for different technologies, i.e., the natural gas, the biomass, the wind, the solar and the battery, which are estimated from all generator data given in Federal Energy Regulatory Commission (FERC) Form 1 [363] by using the EM algorithm [180].

Table 5.2 O&M cost and fitting model by technology.

Technology	Type	Efficiency Reduction
Natural Gas	GE's Jenbacher J312 GS	$\lambda \sim N(0.0043, 0.0005)$, $\sigma_B = 0.0045, a = 1$ [359]
Biomass	GE's Jenbacher J208 GS	$\lambda \sim N(0.0057, 0.0004)$, $\sigma_B = 0.0052, a = 1$ [360]
Wind	Siemens SWT-2.3	$\lambda \sim N(0.0138, 0.0004)$, $\sigma_B = 0.01, a = 1$ [353]
Solar	Advance API156P-200	$\lambda \sim N(0.0048, 0.0005)$, $\sigma_B = 0.004, a = 1$ [361]
Battery	Samsung Prismatic M2967	$\lambda \sim N(0.005, 0.00026)$, $\sigma_B = 0.04, a = 2$ [362]
Technology	Rated Power	O&M Cost ($/KWh)
Natural Gas	400 kW	$\bar{\lambda} \sim N(0.000027, 0.000003)$, $b = 1, \bar{\sigma}_B = 0.000002$, $\bar{X}(0) \sim N(0.0287, 0.00004)$
Biomass	300 kW	$\bar{\lambda} \sim N(0.00009, 0.000003)$, $b = 1, \bar{\sigma}_B = 0.000003$, $\bar{X}(0) \sim N(0.041, 0.0003)$
Wind	200 kW	$\bar{\lambda} \sim N(0.00085, 0.000023)$, $b = 1, \bar{\sigma}_B = 0.00001$, $\bar{X}(0) \sim N(0.005, 0.0001)$
Solar	200 W	$\bar{\lambda} \sim N(0.00002, 0.000002)$, $b = 1, \bar{\sigma}_B = 0.0000012$, $\bar{X}(0) \sim N(0.00918, 0.00002)$
Battery	2 kWh	$\bar{\lambda} \sim N(0.00019, 0.000005)$, $b = 1, \bar{\sigma}_B = 0.0001$, $\bar{X}(0) \sim N(0.0077, 0.0001)$

5.3.2.3 Quantitative Aging Impact

The impact of the aging process on the DGS is measured by the two indicators, i.e., the O&M costs and the energy not supplied, defined by (4.25) and (4.28).

Table 5.3 demonstrates the evolution of indicator Co^s of the DGS over 12 consecutive scenarios s, with respect to three different periods, i.e., slack, flat and peak demand period for the case with and without degradation. These three demand periods help characterizing impact of degradation on the DGS under different operating conditions.

Table 5.3 The indicators $|Co|$ of the DGS over 12 scenarios on September 6, 2013.

Scenario	Slack		Flat
	No Degradation	Degradation (Cost Increase)	No Degradation
1	239.63	15.38	344.64
2	242.86	14.61	348.09
3	237.47	18.37	337.04
4	219.39	17.32	319.34
5	221.21	18.73	323.11
6	217.47	18.9	311.70
7	227.40	17.95	305.04
8	227.55	17.37	306.66
9	234.74	19.8	296.59
10	243.76	18.39	294.43
11	247.27	20.48	289.95
12	255.52	17.41	294.84

Scenario	Flat	Peak	
	Degradation (Cost Increase)	No Degradation	Degradation (Cost Increase)
1	21.62	504.25	29.97
2	21.99	480.06	27.86
3	22.7	466.38	27
4	20.88	422.86	2.96
5	22.28	440.09	18.2
6	18.38	436.31	24.44
7	18.8	421.02	19.17
8	18.2	425.23	30.11
9	22.38	409.17	22.39
10	22.46	392.98	26.25
11	17.48	376.52	19.68
12	21.03	365.07	22.97
$|TCo|$	21.62	504.25	29.97

Table 5.3 shows that the aging process does increase the O&M cost Co^s of every scenario s, by reducing the generators efficiency. The total O&M cost of the peak demand period is increased by 291 $ and it exceeds the increase for the flat demand period 248 and for the slack demand period 234. This is expected because the power generation is bigger during the peak demand period and, given the same level of degradation, also the overall costs are larger.

To minimize the uncertainties introduced by renewable energy resources and loads and have a better estimate of the impact of aging process on the DGS, the MCS can be modified a bit by including more samples and estimating the grand mean.

The grand mean is the mean of the means of several subsamples, as long as the subsamples have the same number of data points [364]. Therefore, the impact of degradation on the DGS from a statistical perspective is described by the mean value of TCo_N, $TENS_N$, TCo_D and $TENS_D$, which can be estimated using 200 MCS runs [364].

Each MCS run contains 500 samples where each sample contains 12 scenarios of DGS from Sept. 2011 to Sept. 2017. The $TCo_{N,j}$, $TENS_{N,j}$, $TCo_{D,j}$ and $TENS_{D,j}$ are the total O&M costs and ENS of the 12 scenarios, i.e., computed by using (4.27) and (4.28), regarded as the outputted indicators of one sample.

Therefore, the estimated TCo_N, $TENS_N$, TCo_D and $TENS_D$ of each MCS run are given as

$$TCo_N = \frac{\sum_{j=1}^{500} TCo_{N,j}}{500}$$

$$TENS_N = \frac{\sum_{j=1}^{500} TENS_{N,j}}{500}$$

$$TCo_D = \frac{\sum_{j=1}^{500} TCo_{D,j}}{500}$$

$$TENS_D = \frac{\sum_{j=1}^{500} TENS_{D,j}}{500}.$$

Finally, by taking the average value of TCo_N, $TENS_N$, TCo_D and $TENS_D$ of all MCS runs, the grand mean can be obtained.

Table 5.4 shows the statistics of the estimated TCo_N, $TENS_N$, TCo_D and $TENS_D$ for 200 MCS runs, i.e., grand mean, and identifies in which period the DGS is most affected by the aging process, i.e., the slack, flat and peak demand period.

The comparison among TCo_N, $TENS_N$, TCo_D and $TENS_D$ shows that the aging process increases the energy not supplied and O&M cost of the DGS. Since the aging process reduces the generators efficiency, the O&M costs to produce the same amount of energy increase, and, therefore, the Total O&M Cost (TCo) of the DGS at the peak demand period increases most when considering degradation.

Because the aging process can reduce generator's maximum capacity, the generator produces less power than its nominal rated power. This means that degradation has caused the demand and supply not to balance by only changing the generators' setpoints, and, therefore, the DGS suffers from large energy not supplied.

As a result, as the peak demand can easily cause many generators to operate close to their reduced maximum capacity, the DGS cannot reserve enough power to meet the peak demand and therefore has larger energy not supplied, compared to the other two demand

Table 5.4 The estimated *TCo* and total ENS (*TENS*) of the DGS of 12 scenarios from September 2011 to September 2017.

Period	All	Slack	Flat	Peak
μ_{TENS_N}	10.33	7.99	10.51	12.51
σ_{TENS_N}	4.27	3.05	3.83	4.56
95% Confidence Interval	[9.99,10.68] [4.05,4.53]	[7.56,8.41] [2.77,3.38]	[9.98,11.04] [3.49,4.25]	[11.87,13.14] [4.15,5.05]
μ_{TENS_D}	10.87	8.14	10.88	13.57
σ_{TENS_D}	4.59	3.06	3.92	4.89
95% Confidence Interval	[10.50,11.23] [4.34,4.86]	[7.71,8.57] [2.78,3.39]	[10.34,11.43] [3.57,4.34]	[12.89,14.25] [4.45,5.41]
$\left\|1 - \dfrac{\mu_{TENS_D}}{\mu_{TENS_N}}\right\|$	5.23%	1.88%	3.52%	8.47%
μ_{TCo_N}	4296	3290	4383	5215
σ_{TCo_N}	790	40.00	51.61	61.42
95% Confidence Interval	[4233,4359] [748,838]	[3285,3296] [36.49,44.45]	[4376,4390] [47.00,57.23]	[5207,5224] [55.93,68.11]
μ_{TCo_D}	4417	3324	4493	5435
σ_{TCo_D}	866	43.02	56.39	68.16
95% Confidence Interval	[4348,4486] [819,918]	[3318,3329] [39.53,46.28]	[4486,4500] [51.80,61.99]	[5426,5443] [62.61,74.93]
$\left\|1 - \dfrac{\mu_{TCo_D}}{\mu_{TCo_N}}\right\|$	2.82%	1.03%	2.51%	4.22%

periods. This explains the results, i.e., the *TENS* of DGSs under the peak demand period is largest, shown by Table 5.4.

A conclusion can be made that the probability of not meeting the demand increases as the demand increases, which results in large *TENS*. Referring to the operators '$|1 - \mu_{TENS_D}/\mu_{TENS_N}|$' and '$|1 - \mu_{TCo_D}/\mu_{TCo_N}|$', the DGS at the peak demand period has largest percentage increment, i.e., 8.47% and 4.22%, in the above two operators, which is consistent with previous discussions.

Figure 5.5 contrasts the probability distributions of TCo_N, $TENS_N$, TCo_D and $TENS_D$ for the statistics in Table 5.4 Column 2 and demonstrates the impact of the aging process on the operations of the DGS.

The aging process has the effect of shifting the distributions to the right, compared to the one without degradation. However, the shift is not extremely pronounced as it is also indicated by the small difference of the indicators $|1 - \mu_{TENS_D}/\mu_{TENS_N}|$' and '$|1 - \mu_{TCo_D}/\mu_{TCo_N}|$ in Table 5.4 Column 2 ("All").

Figure 5.5 also shows that when taking into account the degradation, the occurrence probability of larger *TCo* and *TENS* increases which is consistent with the conclusions made for Table 5.4.

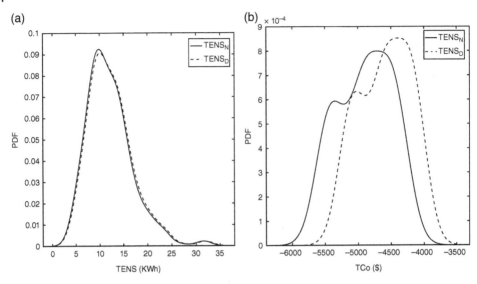

Figure 5.5 The estimated PDF of (a) $TENS_N$ and $TENS_D$, and (b) TCo_N and TCo_D using 200 MCS runs.

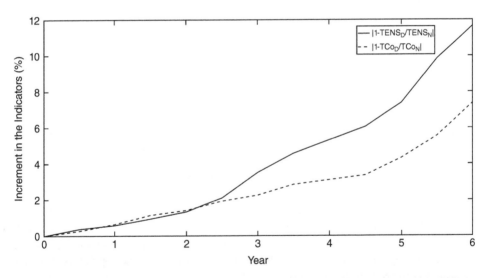

Figure 5.6 The reductions in the two indicators caused by the aging process as a function of time.

Figure 5.6 illustrates the percentage increment in the two indicators, i.e., the TCo and $TENS$, caused by the aging process. As expected, this increment is monotonically increasing with time, and implies the degradation of the generator's efficiency.

The average percentage increments of TCo and $TENS$ in Figure 5.4 are 2.63% and 5.13% over 6 years, respectively, and are consistent with the results in Table 5.4 Column 2.

If the degradation is neglected, the estimates of TCo and $TENS$ at year 6 are around 8% and 12% lower than their real values. This is a large difference which affects the management, maintenance and planning of the DGS to a large extent. The increase of TCo decreases the total DGS revenues and the return on investment.

Additionally, more generation capacity needs to be installed to achieve a given level of power output, as indicated by the increasing *TENS*. Figure 5.6 suggests a slightly lower rate of aging in the first one or two years of the plant's life. This early period is typically covered by comprehensive warranties which guarantee near-maximal output, whereas older plants may be less intensively maintained and thus deteriorate more rapidly, from the third year.

5.4 Applications to Gas Turbine Plant

In the applications to gas turbine plant, the development and optimization of maintenance strategies for degraded control systems is exemplified subject to sudden load increases. The performance features are quantified and the cost analysis of the PBM model is presented. Furthermore, the optimum maintenance strategy obtained from the HGSAA method is presented, and the impact of different combinations of maintenance cost factors on the maintenance strategy is investigated. Last is the comparison studies with other maintenance models to illustrate the advantage of proposed PBM model.

According to [365], the unit maintenance cost, the downtime cost associated with the gas turbine (the time between two consecutive overhauls is 20000~40000 h) and the inspection cost are:

- corrective and preventive maintenance: the expected duration of major overhaul is 45 day and the baseline cost is $c_{PM} = c_{CM} = 5000$ $;
- downtime cost: $c_{DT} = 18000$ $ per 10000 h;
- inspection: the mean duration is 10 d and cost is $c_I = 1600$ $.

5.4.1 Sensitivity Analysis of PBM

5.4.1.1 Impact of Degradation on LFC

Figure 5.7 shows the evolution of the system frequency in two scenarios of different degradation levels, i.e., $T = 0$ (no degradation) and $T = 8$. The aging process increases the maximum transient frequency drop and the frequency recovery time, and thus deteriorates the LFC performance.

The LFC strategy fails to stabilize the frequency at $T = 8$, because the maximum transient frequency drop *PO* exceeds the allowed threshold of -0.016 pu. On the other hand, the frequency recovery time provided by the PID controller is robust against degradation and its value *ST* is well below the maximum allowed M_{ST} in both scenarios. Therefore, preventive maintenance and failure are affected by the maximum transient frequency drop, i.e., M_{PO} and L_{PO}.

5.4.1.2 Numerical Sensitivity Analysis

Based on the maintenance cost factors in Section 5.2.2, the sensitivity analysis of the expected maintenance cycle $E[R_L]$, the average downtime $E[Dt]$, the maintenance cost rate $LC(M, L, T_I)$ and the system availability A with respect to variations of the preventive maintenance thresholds M_{PO} and the inspection interval T_I is performed.

Table 5.5 presents the results of the MCS in terms of M_{PO}, T_I, the percentage of preventive maintenance *PPM* (the ratio between the number of preventive maintenance actions and

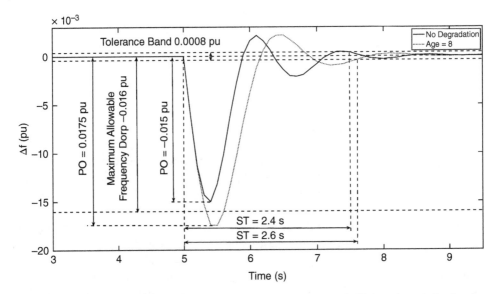

Figure 5.7 The evolution of the system frequency in two scenarios of different degradation levels, i.e., $T = 0$ and 8.

Table 5.5 The predicted PBM features and MCR $LC(M, L, T_I)$ as a function of M_{PO} and T_I.

No.	M_{PO}	T_I	PPM	PCM	$E[R_L]$
1	$15.95 \cdot 10^{-3}$ pu	3.0	0.29	0.61	5.04
2	$15.90 \cdot 10^{-3}$ pu	3.0	1.00	0.00	3.00
3	$15.93 \cdot 10^{-3}$ pu	3.0	0.70	0.15	3.53
4	$15.97 \cdot 10^{-3}$ pu	3.0	0.01	0.49	5.94
5	$15.95 \cdot 10^{-3}$ pu	2.5	0.00	0.50	5.00
6	$15.70 \cdot 10^{-3}$ pu	2.5	1.00	0.00	2.50
7	$15.95 \cdot 10^{-3}$ pu	3.5	0.00	1.00	3.50
8	$15.95 \cdot 10^{-3}$ pu	4	0.00	1.00	4.00

No.	$E[Dt]$	μ_A	σ_A	μ_{LC}	σ_{LC}
1	1.13	62.38%	0.98%	8215	110
2	0.00	100.00%	0	2217	73
3	0.42	86.11%	2.12%	4446	207
4	1.38	54.04%	1.21%	9669	248
5	0.90	64.02%	0.88%	8071	277
6	0.00	100.00%	0	2671	81
7	0.29	91.72%	1.35%	3377	104
8	0.80	80.07%	1.52%	7002	178

the total number of inspections), the percentage of corrective maintenance *PCM* (the ratio between the number of corrective maintenance actions and the total number of inspections), $E[R_L]$, $E[Dt]$, the availability A and the long-term maintenance cost rate $LC(M, L, T_I)$.

Moreover, the simulation results for A and $LC(M, L, T_I)$ are provided with the 95% confidence interval to ensure the prediction accuracy of the proposed maintenance model. The number of samples in the MCS is set to 500. As a result, μ_A and σ_A are the mean and variance of A, and μ_{LC} and σ_{LC} are the mean and variance of $LC(M, L, T_I)$.

The results in Table 5.5 illustrate that the preventive maintenance threshold and the inspection interval of the maintenance strategy have a large influence on the maintenance costs. In particular, decreasing the preventive maintenance threshold on the maximum allowed frequency drop M_{PO}, reduces the expected downtime and reduces $LC(M, L, T_I)$ because the downtime penalty costs are reduced. This is shown by the comparison of case 1, 2, 3 and 4.

Moreover, the expected maintenance cycle $E[R_L]$ is the shortest, if preventive maintenance actions are mainly carried out, and, therefore, the expected downtime is reduced, as shown by case 2. Consequently, the maintenance cost ratio $LC(M, L, T_I)$ is also the smallest for case 2. Finally, the inspection interval T_I is also very critical in determining the MCR, as shown by the comparison of case 1, 5 and 7. If T_I is too short (case 1 and 5), the controller performance does not exceed the preventive maintenance thresholds at the next inspection interval. As a result, the expected maintenance cycle and the system downtime increase, and the availability decreases.

Conversely, if the inspection interval is too long (case 7), the controller performance indicates failure at the inspection, and corrective maintenance is mainly carried out. As a result, the conditions of the case 7 exclusively entail corrective maintenance actions. Case 7 is associated with the smallest $E[R_L]$ and $E[Dt]$, and therefore smallest penalty costs for downtime, as compared to case 1 and 5.

5.4.1.3 Pictorial Sensitivity Analysis

Figure 5.8 shows $LC(M, L, T_I)$ as a function of $M_{PO} \epsilon [-0.016, -0.015]$ pu and $T_I \epsilon [2, 4]$. These ranges cover the failure and the optimum PID condition for M_{PO}, and the usual mean time before maintenance for T_I. If the inspection interval is small, i.e., $T_I = 2.5$, small $|M_{PO}|$ values ensure that the system receives preventive maintenance actions at each inspection.

Moreover, MCR is sensitive to M_{PO} for values $M_{PO} < -15.70 \cdot 10^{-3}$ pu, because the degradation level may not exceed the preventive maintenance threshold. As a result, the system is not maintained at the first inspection, degradation progresses, and the failure threshold is exceeded between two consecutive inspections. This leads to large $E[Dt]$, as the comparison of case 5 and case 6 in Table 5.5 also shows. Therefore, if small inspection intervals are adopted, large $|M_{PO}|$ values are inappropriate for effective preventive maintenance and result in a large MCR.

On the other hand, if the inspection interval is large, i.e., $T_I = 4$, the system performance exceeds the failure thresholds at each inspection and corrective maintenance is carried out. As a result, MCR is not affected by M_{PO}. However, for relatively large T_I, $E[Dt]$ increases as T_I increases because the time between failure occurrence and corrective maintenance

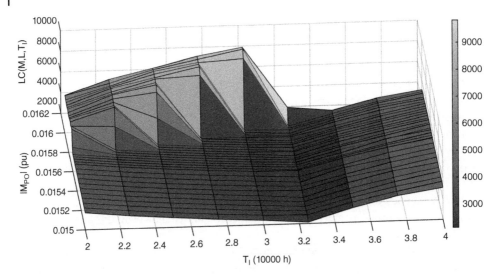

Figure 5.8 The $LC(\boldsymbol{M}, \boldsymbol{L}, T_I)$ as a function of M_{PO} and T_I.

increases as the comparison of case 7 and case 8 in Table 5.5 also shows. Therefore, MCR increases as shown in Figure 5.8.

Finally, Figure 5.8 identifies the combination of M_{PO} and T_I for which the minimum $LC(\boldsymbol{M}, \boldsymbol{L}, T_I)$ is achieved given the specific cost factors. The results of the sensitivity study show the importance of properly designing the preventive maintenance thresholds and the inspection interval, in order to reduce the expected maintenance cycle, improve the system availability and minimize the maintenance cost rate.

5.4.2 Optimal Maintenance Strategy

The MCS-based HGSAA proposed in Section 5.2.2 is applied for identifying the optimal maintenance strategy in the degraded power system for diverse values of the cost factors c_{PM}, c_{CM}, c_{DT} and c_I.

Table 5.6 shows the optimal maintenance strategy $[M_{PO}, T_I]$ in terms of the preventive maintenance threshold and the inspection interval for different combinations of the cost factors $[c_{PM}, c_{CM}, c_{DT}, c_I]$.

The results in Table 5.6 show that different combinations of the cost factors influence the maintenance strategies and result in different levels of the maintenance performance. In particular, the reduction of the inspection cost c_I leads to the reduction of the optimum inspection interval T_I, which in turn increases the observations on the control system and reduces the length of the replacement cycle and the system downtime. This is shown by the comparison of case 1, 2 and 3.

Moreover, high system downtime cost c_{DT} also leads to short inspection intervals T_I and to small preventive maintenance thresholds, which ensure frequent maintenance actions, as shown by case 1, 6 and 7. If the maintenance costs c_{PM} and c_{CM} are high, the optimal maintenance strategy prescribes long inspection intervals, and the preventive maintenance thresh-

Table 5.6 The optimal maintenance strategies $[M_{PO}, T_I]$ for different combinations of the cost factors $[c_{PM}, c_{CM}, c_{DT}, c_I]$.

No.	Cost Factors	Optimal Maintenance Strategy	LC^*
1	[5000, 5000, 18000, 1600]	$[15.79 \cdot 10^{-3}$ pu, 3.14]	2102
2	[5000, 5000, 18000, 1200]	$[15.67 \cdot 10^{-3}$ pu, 3.06]	2028
3	[5000, 5000, 18000, 2000]	$[15.83 \cdot 10^{-3}$ pu, 3.28]	2281
4	[3750, 3750, 18000, 1600]	$[15.63 \cdot 10^{-3}$ pu, 3.10]	1726
5	[6250, 6250, 18000, 1600]	$[15.93 \cdot 10^{-3}$ pu, 3.35]	2574
6	[5000, 5000, 13500, 1600]	$[15.80 \cdot 10^{-3}$ pu, 3.45]	2094
7	[5000, 5000, 22500, 1600]	$[15.71 \cdot 10^{-3}$ pu, 3.01]	2196

No.	$E[R_L]^*$	$E[Dt]^*$	$A^*(\%)$
1	3.14	0.000	100.00
2	3.06	0.000	100.00
3	3.28	0.027	99.19
4	3.10	0.000	100.00
5	3.35	0.043	98.72
6	3.48	0.051	98.53
7	3.01	0.000	100.00

olds approach the failure thresholds, as illustrated by case 5. Under these circumstances, the maintenance actions are less frequent and the system downtime increases, resulting in large maintenance cost rates.

On the contrary, if the maintenance costs are low, the optimal maintenance strategy prescribes short inspection intervals, and the values of the preventive maintenance thresholds are far from those of the failure thresholds, as shown by the case 4. Consequently, the frequency of maintenance actions increases, and the system downtime is reduced. Furthermore, the system availability A^* is strongly influenced by the maintenance cost rate via the average downtime $E[Dt]$, which decreases if the frequency of preventive maintenance actions increases.

Therefore, small maintenance costs and (to some extent) small inspection costs result in large A^* and small $LC^*(M, L, T_I)$, as shown by case 2 and 4. Finally, if the downtime costs decrease, the optimum maintenance strategies lead to large downtime and small A^*. As a result, the inspection interval increases, fewer replacements are carried out, and $LC^*(M, L, T_I)$ is small, as illustrated by case 6.

The proposed MCS-based HGSAA runs on a 64-bit Windows desktop with 32 GB memory and an Intel(R) Xeon(R) E5-1650 v3 @ 3.50GHz CPU. The optimum maintenance strategy is identified in 3.7 days employing 50 generations and a population size of 50 in the GA. This time is acceptable as the MCS-based HGSAA is executed off-line.

5.4.3 Maintenance Models Comparison

The performance-based method is benchmarked against the pre-scheduled and the condition-based maintenance in which the health indexes of the gas turbines determine maintenance activities. If CBM is applied, the turbines are removed for an overhaul when the efficiency and the flow capacity drop by 3% and 6%, respectively [161].

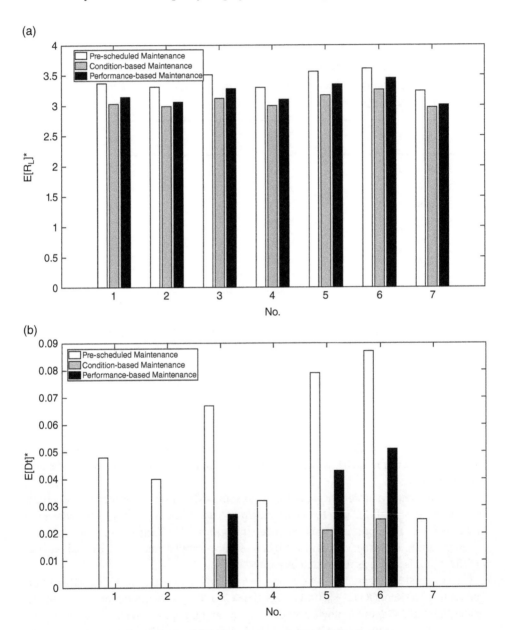

Figure 5.9 Comparison among the pre-scheduled, the condition-based and the performance-based maintenance models after optimization with respect to (a) the length of maintenance cycles $E[R_L]^*$, (b) the system downtime $E[Dt]^*$, (c) the system availability A^* and (d) the long-term maintenance cost rates $LC^*(M, L, T_l)$.

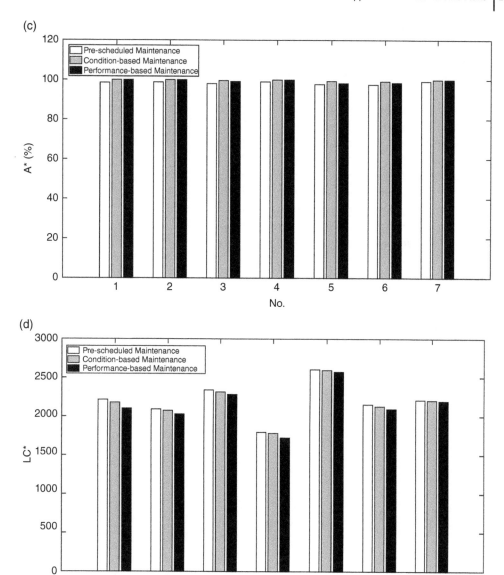

Figure 5.9 (*Continued*)

For the sake of comparability, the MCS-based HGSAA is applied to the pre-scheduled and to the CBM strategy to identify, respectively, the optimal maintenance interval, and the optimal inspection intervals and preventive maintenance thresholds. The comparison is conducted for the seven combinations of cost factors in Table 5.6, and the results of $E[R_L]^*$, $E[Dt]^*$, $LC^*(\boldsymbol{M}, \boldsymbol{L}, T_I)$ and A^* are provided in Figure 5.9.

In the pre-scheduled model, maintenance activities are carried out at fixed intervals. This lack of flexibility results in poor values of the maintenance variables, shown by Figure 5.9. On the other hand, maintenance activities based on the conditions of gas turbines may be conducted even if the LFC performance is still acceptable, if the health indexes of gas turbines have exceeded the failure threshold.

Indeed, the controller compensates the performance loss of the degraded turbines by increasing the load of the healthy ones [157, 176]. Therefore, the CBM has shorter maintenance cycles shown by Figure 5.9 (a), shorter system downtime shown by Figure 5.9 (b) and larger availability shown by Figure 5.9 (c), but results in larger long-term maintenance cost rates as compared to the PBM. Figure 5.9 (d) shows that the proposed PBM model can further reduce the maintenance costs of the pre-scheduled maintenance and the CBM and has the minimum maintenance costs.

The provided reduction of maintenance costs is comparatively small because the current maintenance models are well designed. Nonetheless, the comparison of the maintenance costs in Figure 5.9 (d) shows that the practitioners should decide their maintenance activities based on the criteria which directly determine the system failures, i.e., the LFC performance of power systems in this chapter.

However, because the proposed PBM model only considers the reduction in the LFC performance, it neglects the information of the real-time health indexes of gas turbines, which are critical in predicting the failure time of the individual gas turbines. A comprehensive maintenance model for control systems should account for both the health indexes of individual components and for the control performance, in order to provide cost-effective and adaptive maintenance activities.

6

Game Theory Based CPS Protection Plan

This chapter studies the optimal defense resource allocation for a CPS subject to cyber-attacks and investigates the influence of the most probable attack time and its estimate by the defender. The general CPS is a collection of cyber-physical units connected in parallel logic each of which is made of one cyber component and a set of physical components connected to it [209]. Such structure represents smart grids, networked control systems and Supervisory Control and Data Acquisition (SCADA) systems [215].

In networked control systems, the actuators are out of control and deliver unexpected performance if control signals cannot be received accurately and timely, e.g., due to denial of service attacks. For simplicity but with no loss of generality, the unavailability of the cyber component results in the loss of control and zero capacity of the physical components in the same cyber-physical unit.

In this chapter, the cyber-attack is modeled as a two-stage min-max game in which the attacker takes action, i.e., deciding the optimal number of components to attack, after the defender to cause maximal damage cost, and the attack time is inferred from a data-driven probability distribution. This chapter assumes one attack over the entire contest. Then, the expected damage costs are evaluated by taking into account the unsupplied demand and the inherent value of the cyber-physical unit.

Two approaches are usually adopted to mitigate the impact of cyber-attacks, namely, allocating redundant cyber-physical units and allocating individual protection of unit against the cyber-attacks. Resource allocation strategies in CPSs identify the optimal fraction of defense resources to use in allocating redundant cyber-physical units and enhancing their protection and reduce the total expected damage cost for a given system demand level.

Because of the inherent properties of two-stage min-max games, a closed-form solution of the optimal strategy is not available. Therefore, this chapter makes use of a derivative-free, meta-heuristic algorithm to provide optimal solutions considering the uncertainty on the most probable attack time and its estimate by defender. The Particle Swarm Optimization (PSO) is applied to search for the optimal resource allocation strategy because it allows to improve a candidate solution with regard to a given measure of quality, i.e., the total expected damage cost, makes few or no assumptions about the problem to be optimized and can search very large spaces of candidate solutions [15, 367].

Part of the contents are from [254] and permission has been obtained to use the contents.

Cyber-Physical Distributed Systems: Modeling, Reliability Analysis and Applications, First Edition.
Huadong Mo, Giovanni Sansavini and Min Xie.
© 2021 John Wiley & Sons Ltd. Published 2021 by John Wiley & Sons Ltd.

6.1 Vulnerability Model for CPSs

The vulnerability model of the cyber component subjected to attacks is defined by a contest between the defender and the attacker [203], which is modelled as a two-stage min-max game. The vulnerability of the cyber component i at time t, i.e., $v_{i,t}$, quantifies the probability that component i is destroyed by the attack at time t [203–206, 209].

In the two-stage min-max game, the defender can allocate resources following two strategies, i.e., 1) to enhance the protection of all existing cyber components and 2) to allocate new redundant cyber components. On the other hand, the attacker targets a set of cyber components based on the expected damage cost as detailed in next section.

The outcome of the contest, i.e., the vulnerability, is determined by the resources available to each player and by the contest intensity, m. The payoff matrix is described by the contest success function, that translates the player's available resources into the success probability [203]. This approach is widely employed for modeling the contest from an axiomatic perspective [204–206].

If $Q(t)$ resources are used to attack component i and $q(t)$ resources are used to protect i at time t, the vulnerability of i at time t is expressed as [203–206, 209]

$$v_{i,t} = \frac{Q(t)^m}{q(t)^m + Q(t)^m} = \left(1 + \left(\frac{q(t)}{Q(t)}\right)^m\right)^{-1} \tag{6.1}$$

where $m \geq 0$ is the intensity of the contest, which quantifies how decisively the players compete in the contest; $m = 0$ gives egalitarian distribution, i.e., each player has equal probability to success; $m = 1$ gives proportional distribution, i.e., the success probability is proportional to the player's resources; $m = \infty$ corresponds to winner-take-all situation, i.e., the player with the largest amount of resources always wins. The value of m is based on expert opinion or historical data [209].

Remark 6.1 if the defender has superiority in resources, i.e., $Q(t) < q(t)$, the vulnerability for the case $m > 1$ is much smaller than the vulnerability for the case $m < 1$. Conversely, if the attacker has superiority in resources, i.e., $Q(t) > q(t)$, the vulnerability for the case $m > 1$ is much larger than the vulnerability for the case $m < 1$. Therefore, the player who has superiority in resources always benefits from $m > 1$.

Example 6.1 The following simple examples provide multiple realizations of the two-stage min-max game, and show the impact of the game and the player parameters, i.e., m and $Q(t)$, $q(t)$, respectively, on the game's output, $v_{i,t}$:

(a) If $m = 0.5$, $q(t) = 1$ and $Q(t) = 1$, $v_{i,t}$ equals to 0.5
(b) If $m = 0.5$, $q(t) = 2$ and $Q(t) = 1$, $v_{i,t}$ equals to 0.4142
(c) If $m = 0.5$, $q(t) = 1$ and $Q(t) = 2$, $v_{i,t}$ equals to 0.5858
(d) If $m = 2$, $q(t) = 1$ and $Q(t) = 1$, $v_{i,t}$ equals to 0.5
(e) If $m = 2$, $q(t) = 2$ and $Q(t) = 1$, $v_{i,t}$ equals to 0.2
(f) If $m = 2$, $q(t) = 1$ and $Q(t) = 2$, $v_{i,t}$ equals to 0.8.

These examples demonstrate *Remark* 6.1 and show that increasing resources on protection if $m > 1$, i.e., from case (d) to case (e), is much more effective in reducing the vulnerability as compared to $m < 1$, i.e., from case (a) to case (b).

6.2 Multi-state Attack-Defence Game

6.2.1 Backgrounds of Game Model for CPSs

The original CPS before the cyber-attacks contains N_0 cyber-physical units. Figure 6.1 (a) illustrates an exemplary CPS, i.e., the electric power system, in which the generators

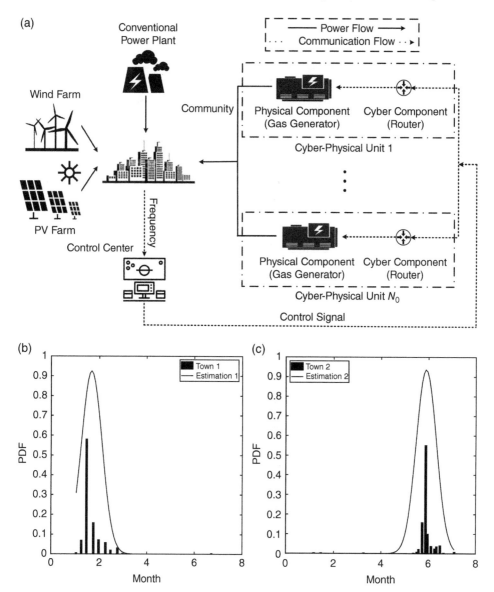

Figure 6.1 (a) Distributed gas generators connected to routers in order to receive frequency control signals are examples of cyber-physical units addressed in this study. (b) The real occurrence time of maximal electricity usage of (b) town 1 and (c) town 2 and their estimations for past 50 years.

directly inject their output into the system and the local demand is not co-located with the power generation. Therefore, only one area is considered, i.e., the community in Figure 6.1 (a) [290].

Diverse technologies are employed to generate power, i.e., conventional power plants, renewable energy resources (wind farms and Photovoltaic Power (PV) farms) and distributed gas generators [220, 221]. In this case, distributed gas generators are actuated by the control center because of their fast response. The N_0 cyber-physical units encompass one physical component (the distributed gas generator) and one cyber component (the router), which receives control signals to regulate the frequency of the power system by balancing generation and demand [194].

If the cyber components are unavailable due to cyber-attacks, e.g., DoS attacks, they cannot receive the control signal from the control center, and, consequently, the physical components connected to them are out of control, disconnected from the power system, and therefore cannot regulate the system frequency. Only one power system area is considered, and therefore tie power lines interconnecting physical components located in different areas are not modelled.

In this configuration, the attack on the cyber component results in the loss of one of the identical distributed gas generators. As a result, the power system in Figure 6.1 (a) can be simplified into a homogenous multi-cyber-physical-unit system. Multi-area configurations of the power systems connected by tie lines may exhibit different impacts with respect to attacks to the cyber elements and to the physical components [234].

Additionally, all units only receive the same individual protection and the overarching protection is not considered in this example [205]. Indeed, the protection strategy studied in this work is a special case of the protection strategies (a mixture of the individual protection and overarching protection) investigated in [192–194, 205].

The attacker chooses to conduct the attack at the occurrence of the maximal demand in the power system [237–239], which is estimated from the historical data in Figure 6.1 (b) and Figure 6.1 (c). The attack can greatly influence the public opinion and likely cause large disruption because it occurs at the maximal demand period [206]. Based on the two-stage min-max single-period game [234], the attacker aims to maximize the system loss by selecting a set of elements to attack.

Conversely, the defender aims to minimize the overall damage cost by deciding optimal resource allocation strategies choosing between the additional redundancy and individual protection. The attack time follows the distribution of the occurrence time of the maximal demand, i.e., it is inferred from its probability distribution [192].

Several assumptions are made in the vulnerability model of the cyber-physical system:

- a cyber-physical unit contains one cyber component and a number of physical components monitored by it;
- the resource stockpiling rates of the attacker and of the defender, namely, A and D, are constant;
- the contest intensity m is constant during the entire contest time T;
- the cost of allocating one redundant cyber-physical unit does not change during the entire contest time T.

For simplicity, two important assumptions are introduced which are common in two-stage min-max games for the risk analysis of complex system [188, 201], namely:

Assumption 6.1 the defender first makes a decision about the resource allocation strategy, and then the attacker decides which components to attack;

Assumption 6.2 the attacker has full information about the defender's strategy and attacks the number of cyber components that maximizes the expected damage cost (see Section 6.4).

6.2.2 Mathematical Game Model

At time t during the contest, the defender has allocated $D \cdot t$ resources with a constant resource stockpiling rate D. Depending on the resource allocation strategy, the rate of resources allocation on the redundancy is $D \cdot r(t)$, where $r(t)$ is the fraction of resource used in installing new cyber-physical units at time t, and $r(t) \leq 1$.

If the cost of allocating one new cyber-physical unit is C, the investigated CPS contains $N_t = \left\lfloor \int_0^t \rho r(\tau) d\tau \right\rfloor$ newly allocated cyber-physical units at time t, where $\rho = D/C$. Therefore, at any time t, the total number of cyber-physical units is $N_0 + N_t$. The amount of resources employed in the protection of the $N_0 + N_t$ cyber components is $\int_0^t D(1 - r(\tau)) d\tau$.

Given the *Assumption 6.1*, the defender has no information on the attacker's resource allocation strategy and, therefore, the protection resources are evenly distributed among all the $N_0 + N_t$ cyber components for individual protection. The protection resource $q(t)$ for each cyber component at time t is

$$q(t) = \frac{\int_0^t D(1 - r(\tau)) d\tau}{N_0 + N_t} \tag{6.2}$$

The attacker can choose to attack any number of $N_0 + N_t$ cyber components, therefore the total number of possible attack strategies is $N_0 + N_t$. Based on the *Assumption 6.2*, the attacker will select the attack strategy which causes maximum expected damage cost at time t.

Assuming that the attacker attacks $n(t)$ cyber components, the attacker distributes the available resources, i.e., $A \cdot t$, evenly across the $n(t)$ attacked cyber components. Therefore, at time t, the attack resource $Q(t)$ devoted to one cyber component among the $n(t)$ targeted cyber components is:

$$Q(t) = \frac{A \cdot t}{n(t)} \tag{6.3}$$

Combining (6.1), (6.2) and (6.3), at time t, the vulnerability model for the cyber component under attack is

$$v_{t|n(t)} = \left(1 + \left(\frac{n(t) \cdot \left(\int_0^t \beta(1 - r(\tau)) d\tau\right)}{t \cdot (N_0 + N_t)}\right)^m\right)^{-1} \tag{6.4}$$

where $\beta = D/A$. Consistently with (6.1), (6.4) quantifies the probability that the cyber component is destroyed if $n(t)$ cyber components are under attack at time t.

Following [368], the developed two-stage min-max game model for cyber-physical systems accounts for:

- system structure: series relationship between cyber and physical component within the same cyber-physical unit (option 2), parallel cyber-physical units (option 3) and multiple elements (option 6);

- defense measure: redundancy (option 3) and protection (option 4);
- attack tactics and circumstances: attack against single (option 1) or multiple elements (option 3), and variable resources (option 7).

6.3 Attack Consequence and Optimal Defence

6.3.1 Damage Cost Model

The expected damage cost model for the CPS in Figure 6.1 (a) is based on the system demand, W, and the cyber component vulnerability model developed in Section 6.2.2.

If the number of cyber components under attack is $n(t)$, the probability that exactly k out of $n(t)$ attacked components are unavailable at time t is

$$p_{k,n(t)}(t) = v_{t|n(t)}^k (1 - v_{t|n(t)})^{n(t)-k} \tag{6.5}$$

where $v_{t|n(t)}$ is the cyber component vulnerability model given in (6.4).

The cost of the unsupplied demand given that k attacked cyber components are unavailable, $\mathcal{L}_k(W)$, depends on the losses associated with the system performance not satisfying the system demand W. If $k \le n(t)$ attacked cyber components are unavailable, the cost of the unsupplied demand is

$$\mathcal{L}_k(W, t) = \varepsilon \cdot max\,(0, W - (N_0 + N_t - k)g) \tag{6.6}$$

where ε is the penalty cost of one unit of unsupplied demand, N_0 is the number of cyber-physical units originally in the CPS, N_t is the total number of newly allocated cyber-physical units by the time t due to the defense strategy, and g is the capacity, i.e., performance, of one cyber-physical unit.

Taking into account the inherent value of the damaged cyber-physical unit and the protection resource deployed on cyber components, the expected damage cost model can be expressed as

$$E_k(W, t) = k(C + q(t)) + \mathcal{L}_k(W, t) \tag{6.7}$$

where C is the inherent value of each cyber-physical unit and $q(t)$ is the protection resource spent on each cyber component.

At time t, under the assumption that the attacker attacks $n(t)$ cyber components, the expected damage cost model subject to system demand W is

$$E(W, t\,|\,n(t), r(t)) = \sum_{k=1}^{n(t)} p_{k,n(t)}(t) E_k(W, t) \tag{6.8}$$

6.3.2 Attack Uncertainty

In realistic applications, the total expected damage cost over the entire contest time T reflects the survival ability of the CPS under cyber-attacks over T. Given the expected damage cost model at time t in (6.8), the total expected damage cost over time T can be computed as

$$E_T(W|n, r) = \int_0^T E(W, t|n(t), r(t))f(t)dt \tag{6.9}$$

The uncertainty about the attack time t, i.e., $f(t)$, heavily impacts the defense strategy. For instance, if the attack time t is known, the defender just needs to find a best resource allocation strategy to minimize damage cost only at this specific time point [234]. On the other hand, if t is unknown, the defender faces a much more complex situation and has to consider allocating resources across the entire contest time [224–226].

Uncertainties related to attack time and defender's estimate on it should be both considered in the optimization problem. In this chapter, the uncertainty regarding to the most probable attack time and its estimate by the defender is represented by the truncated normal distribution. In particular, the attack time t is distributed according to a truncated normal distribution with probability density

$$f(t, a, b, \mu, \sigma) = \frac{\phi\left(\frac{t-\mu}{\sigma}\right)}{\sigma\left[\Phi\left(\frac{b-\mu}{\sigma}\right) - \Phi\left(\frac{a-\mu}{\sigma}\right)\right]}, \quad a \le t \le b \tag{6.10}$$

where ϕ and Φ are the pdf and the cdf of $N(\mu, \sigma^2)$, a is the starting time of the contest, i.e., 0, and b is the contest time horizon T. Thus $\int_0^T f(t, 0, T, \mu, \sigma)dt = 1$ for any μ and σ.

In the context of the two-stage min-max game with attack uncertainty, μ quantifies the most probable attack time, and σ quantifies the accuracy of the defender's estimate on the expected attack time. One example can be found in the Figure 6.1 (b) and Figure 6.1 (c), and other examples of truncated normal distributions can be found in [206, 237].

The optimal defense and attack strategies, i.e., the r^* and n^*, are derived recursively. Given the cyber-attack occurs at time t with probability density $f(t)$, the attacker aims to maximize the damage cost $E(W, t|n(t), r(t))$ in (6.8) by properly selecting $n(t)$.

The attack strategy $n(t)$ depends on the total number of newly allocated cyber components N_t, i.e., $N_0 + \left\lfloor \int_0^t \rho r(\tau)d\tau \right\rfloor$, which is a result of the redundancy allocation fraction $r(t)$.

$$n^*(t) = \arg\max_{n(t)} E(W, t|n(t), r(t))$$

subject to

$$1 \le n(t) \le N_t \tag{6.11}$$

By substituting the optimal $n^*(t)$ into $E_T(W)$ in (6.9), the defender can find the optimal r^* to achieve the minimum total expected damage cost, i.e., $E_T(W|n, r)$, where the dependence on r is made explicit, during the entire contest time T

$$r^*(t) = \arg\min_r E_T(W|n^*, r)$$

subject to

$$0 \le r \le 1 \tag{6.12}$$

Therefore, the attack time t is the free choice variable for the attacker, which is drawn from a data-driven truncated normal distribution. On the contrary, n is chosen strategically by the attacker to maximize the damage cost at time t, and r is chosen strategically by the defender to minimize the damage cost over entire contest time T.

6.3.3 Optimal Defence Plan

The optimization problem defined in above section is non-linear and non-convex. Analytical results can be achieved only in simplified and particular situations. On the other hand, heuristics are usually employed in solving this type of problems. In particular, the PSO is often applied for solving various global optimization problems, especially non-linear and non-convex problems [15, 367].

To this aim, a solution including the defense resource allocation strategy is encoded as a finite-length string called "particle" in PSO. The fitness value of each particle is determined by the fitness function (6.9).

During each iteration of PSO, information about the best position of each particle and the best particle in the entire swarm can be updated after computing the fitness values of all particles. And then in the next iteration, the movements of the particles are guided by their previous best-known positions within the search space, called *pbest*, as well as the best-known position of the entire swarm, called *gbest*. This can avoid that the PSO is trapped into local minima.

For a particle i with j dimensions, its local best known position can be represented by $pbest_i = (p_{i1}, p_{i2}, ..., p_{ij})$ and the global best known position of the entire swarm is $gbest = p_g$. Thus, the velocity v_{ij}^{I+1}, and the position x_{ij}^{I+1}, associated with the dimension j of the particle i after iteration I, are updated according to:

$$v_{ij}^{I+1} = w \cdot v_{ij}^I + c_1 \cdot \rho_1 \cdot (p_{ij} - x_{ij}^I) + c_2 \cdot \rho_2 \cdot (p_g - x_{ij}^I) \tag{6.13}$$

$$x_{ij}^{I+1} = x_{ij}^I + v_{ij}^I \tag{6.14}$$

where $I \in \{1, ..., NI\}$; $i \in \{1, ..., NP\}$; $j \in \{1, ..., J\}$; w is the inertia weight; c_1 and c_2 are the cognition learning factor and the social learning factor; $c_1 + c_2$ generally equals to 4 and ρ_1 and ρ_2 are uniformly random factors restricted to $[0, 1]$, based on [236].

The followings are some important steps of the PSO method:

- **STEP 1:** Randomly initialize the particles representing the defense resource allocation strategies, i.e., the velocity and the position.
- **STEP 2**: Compute the fitness value for each particle based on (6.9).
- **STEP 3**: Update the values of *pbest* and *gbest* based on calculated fitness values; determine the velocity and position values for each particle in the next iteration on the basis of (6.13) and (6.14).
- **STEP 4**: If the stopping condition-the maximum number of iterations NI is reached, go to **STEP 5**; otherwise, go to **STEP 2**.
- **STEP 5**: Decode the particle with the minimum fitness value and take the result as the optimally designed defense resource allocation strategy, which can ensure the minimum total expected damage cost for contest time T.

6.4 Applications to Distributed Generation Systems (DGSs) with Uncertain Cyber-attacks

6.4.1 Settings of Game Model

This section investigates the results of the optimal resource allocation strategy for an attacker-defender contest modelled as a two-stage min-max game.

The influence of the most probable attack time and the defender's estimate on it, on the optimal resource allocation strategy is investigated for both constant and dynamic redundancy allocation fraction. The basic settings for the two-stage min-max one-period game [192, 206], and for the PSO algorithm [15] are presented in Table 6.1.

For the optimal resource allocation strategy, a constant redundancy allocation fraction, i.e., $r(t)$, is common and widely used in many studies [192, 201–204]. However, a dynamic redundancy allocation fraction enhances flexibility in complex contests, i.e., in a two-stage min-max game or a game with false targets, at the expenses of additional computational cost [206, 209].

Therefore, Section 6.4.2 studies the optimal resource allocation strategy with a constant redundancy allocation fraction, which is quantified using the enumeration method (step size of 0.01 and average running time is 276 s), and Section 6.4.2 demonstrates the limitations of this assumption and investigates optimal resource allocation strategy with dynamic redundancy allocation fraction, which is quantified using the PSO algorithm.

6.4.2 Optimal Protection with Constant Resource Allocation

6.4.2.1 Impact Under Constant Case

The influence of the most probable attack time μ on the optimal parameters of the constant resource allocation strategy over the contest period T is studied, when the accuracy of the defender's estimate on the attack time, σ, is constant.

Figure 6.2 presents the minimum total expected damage cost $E_T(W|n, r)$ for different combinations of contest intensity $m = 0.2$ and 2, construction cost of the redundant cyber-physical unit $\rho = 0.2$ and 4, $\beta = 0.2$ and 4, system demand $W = 25$, 30 and 35, and $\sigma = 4$.

Table 6.1 Settings for the Contest and the PSO.

Parameter	Value	Parameter	Value
Contest period T	20	Number of particles NP	50
Initial unis N_0	3	Number of dimensions J	10
Capacity g	5	Maximum velocity v	1
Penalty cost ε	2	Maximum position x	1
Inertia weight w	0.6	Cognitive factor c_1	2
Iterations of PSO NI	120	Social factor c_2	2
Resource stockpiling D	20		

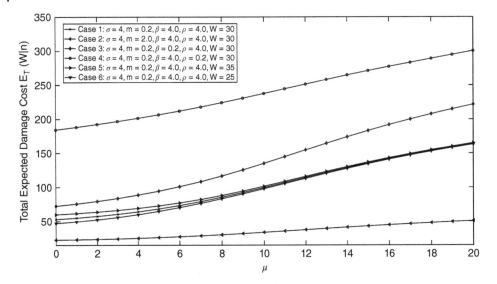

Figure 6.2 Total expected damage cost as a function of the most probable attack time.

The particular choice of σ is consistent with the investigated range of the most probable attack time $\mu \in [0, 20]$, and captures the fact that the defender contemplates a plausible range of attack time before or after the most probable attack time. Furthermore, the selection of σ affects the scale of the results but not their overall trend.

For all cases, $E_T(W|n, r)$ increases for increasing μ because the attacker has an extended amount of time to stockpile the attack resources. Due to Assumption 6.1 and 6.2, the attacker always selects the optimum number of cyber components to attack and causes the maximum damage. On the other hand, the defender is unaware of the attacker's strategy and evenly distributes the defense resources on the protection of each component.

Therefore, the defender is always at disadvantage and can only make an optimal resource allocation strategy under information inequality. This is in line with the intuition that the attacker can make the most efficient use of the increased resource as compared to the defender. This explains the increase of $E_T(W|n, r)$ even if the defender has also an extended amount of time to stockpile the defense resources.

The expected system unsupplied demand as a function of the most probable attack time μ is presented in Figure 6.3 for four combinations of contest intensity $m = 0.2$ and 2, construction cost of the redundant cyber-physical unit $\rho = 0.2$ and 4, and system demand $W = 25$, 30 and 35.

Case 4, 5 and 6 show that the attacker benefits from a contest characterized by intensity $m < 1$, even if the defender has superiority in resources, i.e., $\beta = 4$ (this is consistent with *Remark 1*). Therefore, the expected unsupplied demand increases as the expected attack time μ increases.

The differences in the expected unsupplied demand for Case 4, 5 and 6 stem from the different system demand W. Case 6 has the smallest system demand $W = 25$ and entails the smallest unsupplied demand. Furthermore, the resource allocation by the defender increases as the attack time is expected at a later stage, i.e., increasing μ, and, therefore, the total system value increases and the attack creates larger damage costs.

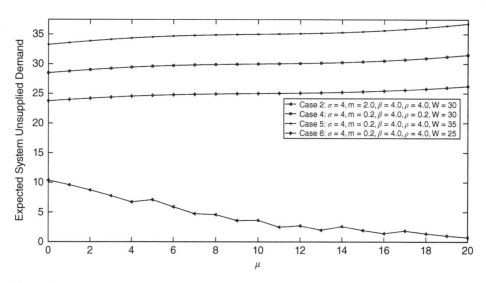

Figure 6.3 Expected system unsupplied demand as a function of the most probable attack time.

The comparison between Case 4 and Case 5 in Figure 6.2 and Figure 6.3 shows that if the construction cost of a redundant cyber-physical unit is large, i.e., $\rho = 0.2$ in Case 4, its inherent value is also large, and, the attack results in large $E_T(W|n, r)$ as compared to Case 5, which has small construction costs, i.e., $\rho = 4$, even if Case 5 entails a large unsupplied demand.

Finally, Case 2 in Figure 6.3 shows that an extended time for protecting and allocating redundancies allows the defender to mitigate the effects of information inequality, if the defender has superiority in resource ($\beta = 4$) and $m > 1$. Therefore, $E_T(W|n, r)$ has a slow growth trend in Figure 6.2 and the expected unsupplied demand decreases as the expected attack time μ increases, and defense resources are accumulated.

6.4.2.2 Optimal Constant Resource Allocation Fraction

Figure 6.4 shows the optimal fraction r^* as a function of the most probable attack time μ for the same scenarios presented in Figure 6.2. The optimal fraction r^* has an overall decreasing trend as a function of μ, i.e., as the attack is likely to occur at a later stage of the contest.

In such case, the defender can allocate enough redundant cyber-physical units before the attack, and the defense resources can be used for enhancing the protection of cyber components. Case 4 shows an abrupt rise of the optimal fraction r^* at the beginning of the contest. Indeed, if an early attack is expected, i.e., $\mu = 0$, and the cost of allocating one redundant cyber-physical unit is large, i.e., $\rho = 0.2$, the defender is urged to allocate enough redundant cyber-physical units in a short time, which results in large r^*.

Case 2 in Figure 6.4 shows that, according to *Remark 6.1*, the defender who has superiority in resources, i.e., $\beta = 4$, benefits from $m > 1$, and substantially limits $E_T(W|n, r)$. In such case, the defender prefers to spend most of resources in protection rather than redundancy allocation, which can be proven by (6.4), as shown in Figure 6.4.

Redundancy plays an important role in satisfying the system demand W: in Case 4, large redundancy allocation costs lead to a large $E_T(W|n, r)$. In Case 1, 5 and 6, a large system

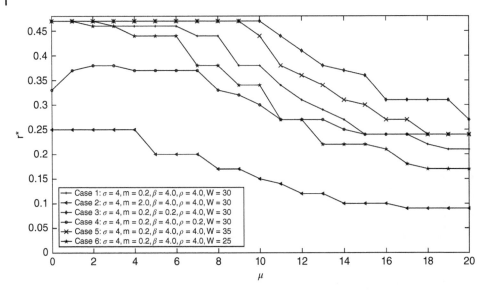

Figure 6.4 Optimal redundancy-allocation rate r^* as a function of the most probable attack time.

demand requires a large r^* to allocate redundant cyber-physical units to provide increased system performance.

Figure 6.5 shows the optimal number of cyber components n^* that the attacker should target, to cause the maximum expected damage $E(W, t|n(t), r(t))$ as a function of time t, over the contest time $T = 20$, and for different combinations of fixed redundancy fraction $r = 0.15$, 0.17, 0.32 and 0.34, the most probable attack time $\mu = 8$ and 10, defender's estimate $\sigma = 2$, 4 and 6, contest intensity $m = 0.2$ and 2, construction cost of the redundant cyber-physical unit $\rho = 0.2$ and 4, $\beta = 4$, and system demand $W = 25$ and 30.

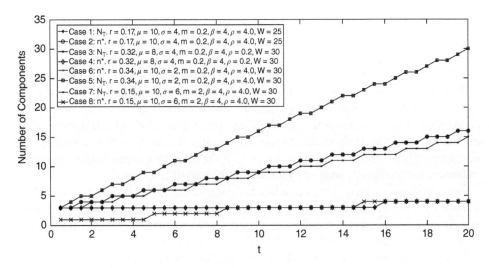

Figure 6.5 Changing curve of optimal number of attacked cyber component n.

For each case, the total number of cyber-physical units in the CPS, i.e., $N_T = N_0 + N_t$, is also reported. At each time t, the attacker determines optimal $n^*(t)$ by maximizing the expected damage $E(W, t|n(t), r(t))$ in (6.8) given a fixed redundancy fraction r.

Based on (6.9), the optimal $n^*(t)$ also leads to a maximum $E_T(W|n, r)$. The selected values cover the significant range of redundancy allocation fractions, attack time, defender's estimate on the attack time and contest intensity. r and $n^*(t)$ are mutually dependent in the two-stage min-max game. Additionally, different costs for deploying redundant cyber-physical units and system demands are considered, which impact on the optimum $n^*(t)$.

Based on (6.4), m is critical in the determination of $n^*(t)$. If $m > 1$ and the defender has superiority in resources, i.e., $\beta = 4$, the attacker targets a subset of the cyber components to maximize the expected damage; thus, $n^*(t) < N_T$ as shown in Case 7 and 8. Conversely, in Case 1 and 6, where $m < 1$ and the defender has more resources, i.e., $\beta = 4$, the attacker targets all cyber components. Indeed, if $m < 1$ the less resourceful contester (the attacker) is favored based on *Remark 1*, and the vulnerability increases. Based on the discussions for (6.5), allocating more resource on protection in case $m < 1$ cannot effectively reduce the vulnerability compared to case $m > 1$, thereby the importance of enhancing protection decreases. Therefore, $n^*(t) = N_T$ for Case 1 and 6.

Figure 6.6 shows the influence of the defender's accuracy on the estimate of the most probable attack time, i.e., σ, on the optimal parameters of the constant resource allocation strategy for $\mu = 10$.

As σ increases, the defender is unable to predict the attack time accurately, the efficiency of his/her resource allocation strategy is disrupted, and thereby the expected damage cost $E_T(W|n, r)$ monotonically increases as shown in Figure 6.6. As σ inceases, $E_T(W|n, r)$ converges the largest total expected damage cost of the completely uncertain case, i.e., when the attack time is modelled as a uniform distribution between 0 and the contest time $T = 20$.

The optimal redundancy fraction r^* has an overall upward trend except for Case 4. If the defender has superiority in resources, i.e., $\beta = 4$, it just conservatively provides more capacity against the increasing uncertainty by enhancing redundancy allocation. However, in Case 4, the defender reduces the redundancy allocation fraction and spends more resource on protection due to high construction costs, i.e., $\rho = 0.2$.

In Case 5, r^* shows an abrupt drop for increasing σ, because the defender becomes unsure on the most probable attack time and is trapped in a dilemma whether to keep allocating the same level of resources on redundancy. As system demand W increases, the defender prefers to slightly increase the resource on redundancy, see the differences between Case 1, 5 and 6. For the smallest demand, i.e., $W = 25$ in Case 6, the drop in r^* disappears because the same strategy is able to satisfy a small demand.

6.4.3 Optimal Protection with Dynamic Resource Allocation

6.4.3.1 Vulnerability Model Under Dynamic Case

Section 6.4.2 shows that a defensive strategy with constant redundancy-allocation fraction is conservative and cannot provide enough flexibility in the dynamic two-stage min-max game, therefore leading to large expected damage costs.

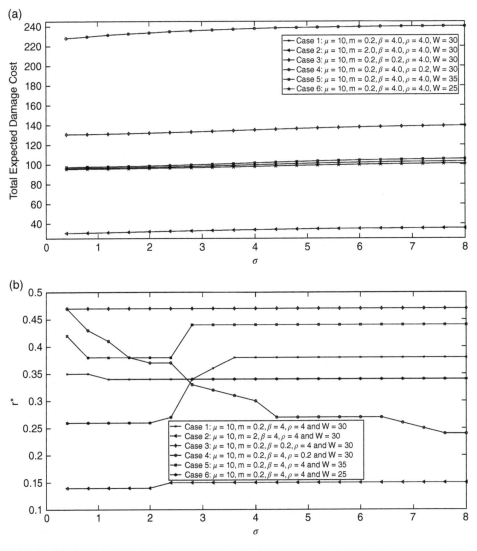

Figure 6.6 Simulation results of (a) total expected damage cost and (b) optimal constant resource allocation as a function of defender's accuracy on the estimate of the most probable attack time, i.e., σ, for $\mu = 10$.

For instance, when the contest intensity is high and the construction cost of redundant cyber-physical unit is low, i.e., Case 2 in Figure 6.2, the protection of cyber components is more important at the early stage of the contest, whereas the redundancy allocation is more important at the later stage of the contest. Therefore, it is reasonable to adopt a small redundancy-allocation fraction at the early stage of the contest, and increase it at a later stage, as compared to adopting a constant one. Thus, a dynamic strategy for the redundancy-allocation fraction $r(t)$, is more appropriate and is investigated in this subsection.

The dynamic redundancy allocation fraction, i.e., $r(t)$, is described by a widely used geometric progression model. It allows the defender to adjust the defense resource allocation strategy by varying $r(t)$.

The geometric progression can describe the dynamic properties of many realistic applications, e.g., the dynamic defense resource allocation strategy [206], the event-altered rate model in software reliability analysis [369] and the topology configuration of inverters to maximize the number of voltage levels from Direct Current (DC) sources [370]. At time t, the construction rate fraction r_j of constructing the j-th cyber-physical unit is given as

$$r_j = \begin{cases} r_1 z^{j-1}, & \text{if } r_1 z^{j-1} < 1 \\ 1, & \text{otherwise} \end{cases} \tag{6.15}$$

where the scale factor r_1 ($0 \leq r_1 \leq 1$) is the construction rate of the first redundant cyber-physical unit, and the common ratio z ($z \geq 0$) determines the evolution of the construction rate fraction.

The construction of j-th redundant cyber-physical unit is completed by time F_j:

$$F_j = \sum_{k=1}^{j} 1/(\rho \cdot r_k) \tag{6.16}$$

where $F_0 = 0$ by definition.

By time t, the number of allocated redundant cyber-physical units N_t is

$$\sum_{k=0}^{N_t} F_k < t < \sum_{k=0}^{N_t+1} F_k \tag{6.17}$$

As indicated by [206], it is reasonable to take into account the cost associated with each redundancy-allocation-rate change, which is the only observed variable in the dynamic resource allocation strategy. If the constant and dynamic resource allocation strategies are compared solely based on the unit construction costs, then the latter is always preferable to the former.

Indeed, the dynamic strategy entails more flexibility and can provide a better or, at worst, equal allocation of resources as the constant strategy. Therefore, including the penalty cost for varying the redundancy-allocation rate results in different construction costs and allows for comparing the two strategies in a consistent way. The penalty cost is a function of the number of changes in the redundancy-allocation rate. At time t, the number of construction rate changes M_t is

$$M_t = \begin{cases} 0, & \text{if } N_t = 1 \\ N_t, & \text{if } r_1 z^{N_t-1} \neq r_1 z^{N_t} \text{ and } N_t > 1 \\ M_0 - 1, & \text{otherwise} \end{cases} \tag{6.18}$$

where M_0 is the smallest integer which guarantees the inequality $r_1 z^{M_0-1} \geq 1$ when $z > 1$. The construction rate fraction change is counted starting from the second allocated redundant cyber-physical unit.

Thus, at time t, the penalty cost model adopted in this work is defined as

$$Cost_t = l \cdot M_t^2 \cdot C \tag{6.19}$$

where l is a coefficient proportional to the construction cost of one cyber-physical unit and quantifies the impact of the dynamic resource allocation strategy on the construction cost. The assumed penalty cost function is convex and increases with the number of redundancy-allocation-rate changes.

As the geometric construction rate fraction strategy consumes part of the defender's resources, the resources used for protection and redundancy decrease as the contest time progresses. Therefore, (6.2) and (6.3) are modified as

$$
\begin{cases}
q(t) = \dfrac{D \cdot t - \sum_{j=1}^{N_t} r_j \cdot (F_j - F_{j-1}) \cdot D - r_{N_t+1} \cdot D \cdot (t - F_{N_t}) - l \cdot M_t^2 \cdot C}{N_t + N_0} \\
Q(t) = \dfrac{A \cdot t}{n(t)}
\end{cases}
\tag{6.20}
$$

and the vulnerability model of the attacked cyber component at time t in (6.4) is modified as

$$
v_{t|n(t)}(r_1, z) = \left(1 + \left(\beta \left(\frac{t - \frac{N_t + l \cdot M_t^2}{p} - r_{N_t+1} \cdot (t - F_{N_t})}{t} \right) \frac{n(t)}{N_t + N_0} \right)^m \right)^{-1}
\tag{6.21}
$$

Similar to the constant case, the attack time t is also drawn probabilistically from the data-driven truncated normal distribution and the number of elements which the attacker should target, i.e., $n(t)$, is chosen strategically as the results of a two-stage min-max game. Conversely, in the constant case, the fraction r for the defense resource allocation is chosen strategically, while in the dynamic case both the construction rate of the first redundant cyber-physical unit, i.e., r_1, and z are chosen strategically.

6.4.3.2 Optimal Dynamic Resource Allocation Fraction

Figure 6.7 shows the influence of the most probable attack time μ on the optimal parameters of the dynamic resource allocation strategy, i.e., $E_T(W|n, r)$, r_1^*, and z^*. Combinations of contest intensity $m = 0.2$, construction cost of the redundant cyber-physical unit $\rho = 4$, $\beta = 4$, system demand $W = 25, 30$ and 35, and $\sigma = 4$ are selected for representing significant scenarios, i.e., the defender has superiority in resource ($\beta = 4$); the construction cost is very small ($\rho = 4$); the attacker can get benefit from the low-intensity contest ($m = 0.2$).

Case 1 and 2 present the results for the dynamic and constant strategy, respectively, and show that the dynamic resource allocation strategy outperforms the constant one and achieves a smaller $E_T(W|n, r)$. For small μ, i.e., the attack would most probably take place at the early stage of the contest, the dynamic strategy prioritizes the allocation of resources on redundant cyber-physical units, i.e., construction rate of each cyber-physical unit ($r_1^* \cdot (z^*)^{j-1}$) is relatively large for small μ and decreases for increasing μ.

In Case 1, 3 and 4, the defender allocates a smaller portion of resources to redundancy as μ increases and z^* is smaller than 1. Though $m < 1$ leads to priority of redundancy, large μ allows enough time to the defender to allocate enough redundant cyber-physical units. Therefore, increasing redundancy is not efficient for large μ, considering that the attacker also gathers more resource and always selects optimal number of attacked cyber components n. Therefore, the defender firstly keeps a high fraction for redundancy and then reduces it to enhance the resource allocation to cyber-unit protection.

Figure 6.8 illustrates the influence of defender's estimate on most probable attack time on optimal parameters of dynamic resource allocation strategy for different combinations

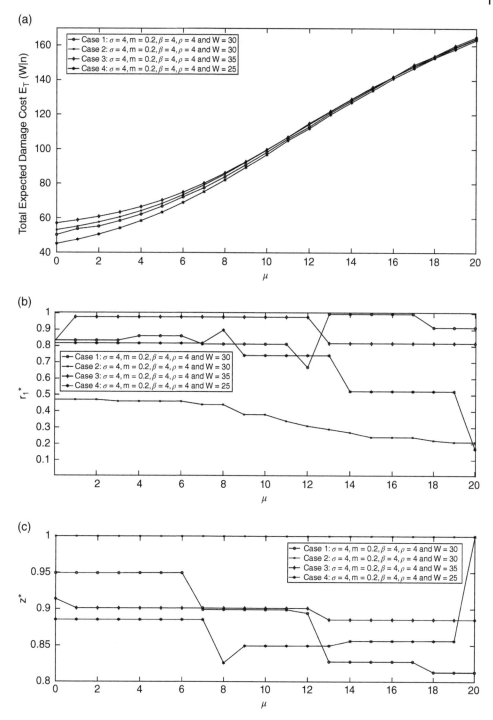

Figure 6.7 Simulation results of (a) total expected damage cost, and optimal dynamic resource allocation (b) r_1^* and (c) z^* as a function of the most probable attack time.

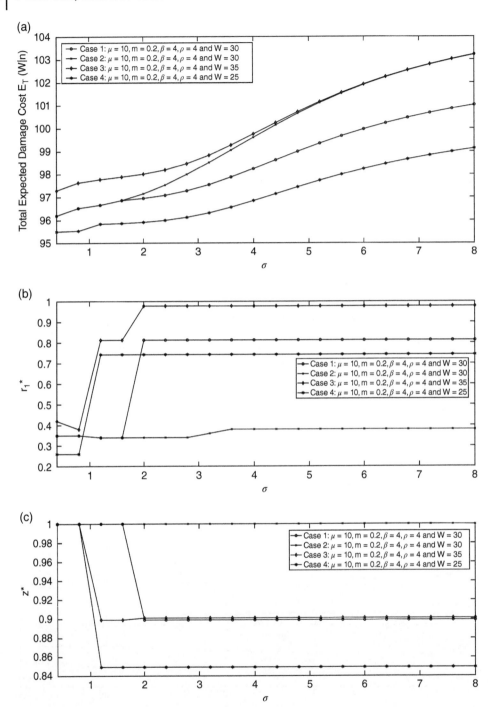

Figure 6.8 Simulation results of (a) total expected damage cost, and optimal dynamic resource allocation (b) r_1^* and (c) z^*, as a function of the accuracy on defender's estimate about the expected attack time.

of contest intensity $m = 0.2$, construction cost of the redundant cyber-physical unit $\rho = 4$, $\beta = 4$, system demand $W = 25$, and 30, and $\mu = 10$.

Case 1 and 2 present the results for the dynamic and constant strategy, respectively, and show that the dynamic resource allocation strategy outperforms the constant one and achieves a smaller $E_T(W|n, r)$, for different levels of defender's accuracy on the estimate of the most probable attack time, i.e., σ. If defender is rather certain on the attack time, i.e., σ is small, the dynamic and constant strategy lead to similar resource allocations as seen in comparing Case 1 and 2. The reason is that the penalty cost for changing the redundancy-allocation rate consumes parts of the defender's resources, and the dynamic strategy is inefficient if the defender is certain about the attack time.

On the contrary, for Case 1, 3 and 4, when uncertainty on the most probable attack time increases, defender has no information on attack time and conservatively keeps a constant strategy. The reason for z^* smaller than 1 is similar to the discussion provided for Figure 6.7.

6.4.3.3 Optimization Results Justification

The application of PSO is justified by two comparison studies among the PSO, the Genetic Algorithms (GA) and the fmincon function in Matlab. The number of particles and number of iterations in PSO are 50. The population size and the generations in GA are 50. In addition, the initial particle position of the PSO and the initial population of the GA should contain the same elements to ensure comparability.

The result of the fmincon depends on the starting point and in the local minimum where the algorithm is trapped, therefore, we build samples of 50 fmincon evaluations using the same starting points as for the PSO and the GA. Therefore, the initial conditions are the same for the three methods.

The comparison is based on 100 samples of solutions for each method under different combinations of μ, σ, m, ρ, β and W. We use the results of the PSO as the base value to calculate the percentage difference ($\frac{E_{T,1} - E_{T,2}}{E_{T,1} + E_{T,2}} \times 100$, where $E_{T,1}$ and $E_{T,2}$ are the total expected damage cost of the fmincon or the GA, and of the PSO, respectively).

The average percentage difference between PSO and the other two methods, i.e., the fmincon and the GA, is 0.20% and 0.0087%, respectively; the average running time for the PSO, the GA and the fmincon is 1153 s, 930 s and 353 s. Therefore, the PSO identifies the smallest total expected damage cost and entails the longest running time. This is due to its large search space stemming from the information exchange among neighboring swarms.

Moreover, the PSO can still identify the smallest total expected damage cost, even when the running time of the three methods is close. The average percentage difference between the fmincon or the GA and the PSO is 0.037% and 0.0035% for the average running time of 1182 s (150 fmincon samples) and 1138 s (population size 60 for GA). Therefore, PSO is used in this work.

The fmincon is easily trapped into local minima (due to the lack of information exchange among neighboring solutions), whereas PSO's domain of search space is large and it can use both local and global information to avoid local minima. In addition, the update process for the particle position in the PSO is more effective than the generation of the new population in the GA.

7

Bayesian Based Cyberteam Deployment

This chapter studies the cybersecurity of smart grids that are the most critical applications of CPSs and has become one of key problems in developing reliable modern power and energy systems.

This chapter quantifies a non-stationary adversarial cost with a variation constraint for smart grids and enables stakeholders to investigate the problem of optimal smart grid protection, i.e., optimal cyberteam deployment, against cyber-attacks in a relatively practical scenario. In particular, a Bayesian Multi-Node Bandit (MNB) model with adversarial costs is developed and a new regret function is defined for this model. An algorithm - Thompson-Hedge algorithm is introduced to solve the problem and the superior performance of the proposed algorithm is proven in terms of the convergence rate of the regret function.

The applicability of the algorithm to real smart grid scenarios is verified and the performance of the algorithm is also demonstrated by realistic examples from Elia Grid, Belgium.

7.1 Poisson Distribution based Cyber-attacks

7.1.1 Impacts of DoS Attack

In the electricity network with cyber-attacks, an attacker launches a coordinated attack using multiple attack vectors, because the smart grid communication networks are physically distributed and highly heterogeneous.

The successful rate of such attack behavior is data-driven, which follows a Poisson distribution, as demonstrated by the empirical study from U.S. Department of Energy [259]. The network defender aims to optimally allocate cyber defense teams among nodes in the network, via probing one node per day.

Such a defending action thwarts all attempted cyber-attacks to that node on that day, and also helps update his/her belief about the uncertain successful rates of attack. The above interaction between the attacker and the defender has the sequential decision-making nature and leads itself to a Bayesian MNB model. This model employs proactively defense teams that traditionally respond to cyber threats after they occur.

Part of the contents are from [375] and permission has been obtained to use the contents.

Cyber-Physical Distributed Systems: Modeling, Reliability Analysis and Applications, First Edition.
Huadong Mo, Giovanni Sansavini and Min Xie.
© 2021 John Wiley & Sons Ltd. Published 2021 by John Wiley & Sons Ltd.

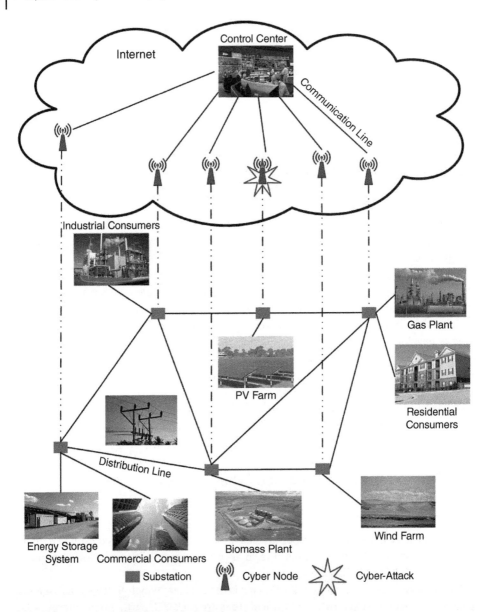

Figure 7.1 The cyber-physical structure of smart grid.

The considered attack scenario in a realistic smart grid as shown in Figure 7.1 is that the Denial of Service (DoS) attacks block the demand response (DR) messages and dispatch commands [259, 371]. The DoS attacks can be accomplished by flooding the communication channel, e.g., the one between the demand response automation server and customer systems or the one between the control center and power plant router, with other messages and commands, or by tampering with the communication channel [259, 371, 372].

Above actions can prevent legitimate DR messages and dispatch commands being received and transmitted, i.e., depriving authorized access or control to customer systems and power plant, resulting in demand not being responded and power plant not being controlled [371]. Therefore, the impact of successful DoS attacks on the optimal power flow model of the smart grid is described by making the target node temporarily unavailable and disconnected from the grid, which is illustrated in Figure 4.12.

7.1.2 Poisson Arrival Model Verification

The interarrival time of cyber-attacks are modelled by an exponential distribution [259] as

$$P(t) = \lambda e^{-\lambda t} \tag{7.1}$$

where interarrival time from 3 to 6 days is lively overrepresented because incidents are only recorded on weekdays.

This section conducts the Chi-square goodness-of-fit test on the cyber incidents from HACKMAGEDDON. Note that if the average interarrival time is $\bar{t}$ with reciprocal rate value $\bar{\lambda}$. To use a Chi-square goodness-of-fit test, form a hypothesis as follows:

- Null hypothesis H_0: the random interarrival time follows the exponential distribution
- Alternative hypothesis H_0: the random interarrival time does not follow the exponential distribution

The Chi-square test is proceeded with intervals of equal probability. If $k = 5$ intervals are chosen, the probability of an observation falling into any of them will be $p = 0.2$. since the cumulative distribution of the exponential is given by

$$F(a_i) = 1 - e^{\lambda a_i} \tag{7.2}$$

When a_i is the endpoint of interval i, $i = 1, 2, ..., k$. Dividing the range of the dependent variable into equal parts given $F(a_i) = ip$. Therefore, there exists

$$ip = 1 - e^{\lambda a_i} \tag{7.3}$$

Which can be solved for a_i with the following result:

$$a_i = -(1/\lambda) \ln(1 - ip) \tag{7.4}$$

The results Chi-square goodness-of-fit test about the cyber-attacks at the channel connecting the control center and the power plant from 2010.06 to 2018.09 are shown by Figure 7.2.

$h = 0$ indicates that Chi-square goodness-of-fit test does not reject the null hypothesis at the default 5% significance level and p-value $p = 0.6848$ (>0.05) indicates weak evidence against the null hypothesis, so the test fails to reject the null hypothesis, i.e., it is data supported to select the exponential distribution to model the interarrival of cyber-attacks.

7.1.3 Average Arrival Attacks

Consider a smart grid with N nodes that suffers cyberattacks. Let $\mathbb{N} \triangleq \{1, ..., N\}$ be the set of all nodes.

At each time $t = 1, 2, ...$, the operator's action is to choose a node to probe. If $i \in \mathbb{N}$ is probed, the operator observes a random number of $K_{i,t}$ cyber-attacks on node i.

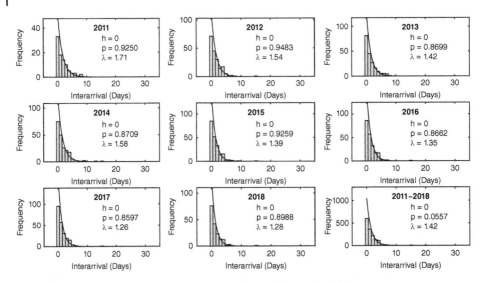

Figure 7.2 Estimation result of cyber-attacks from 2010.06 to 2018.09.

The historical cyber incident data from US Department of Energy (Fig. 2 of [259]) has demonstrated that the arrival interval of successful cyber-attacks on per node can be well described by a truncated Poisson distribution, and there is no record of multiple cyber-attacks (maximum is 3) on per node per day.

Therefore, the physical meaning is that in real applications, there should not be more than 3 cyber-attacks arriving per node in a defined time interval. In addition, the Palm-Khintchine theorem [259, 373] justifies that the aggregate arrivals from many sources (no need to be strict Poisson) approach a truncated Poisson distribution in the limit time interval.

It is assumed that for all $i \in \mathbb{N}$, $\{K_{i,t}, t = 1, 2, \ldots\}$ is a sequence of independent and identically distributed random variable sequence drawn from a Poisson distribution truncated at $m > 0$ with rate λ_i, i.e., for all $k = 1, 2, \ldots$, and $t = 1, 2, \ldots$:

$$\Pr(K_{i,t} = k | \lambda_i) = \begin{cases} \frac{\lambda_i^k}{k!} e^{-\lambda_i}, & k < m \\ \sum_{j=m}^{\infty} \frac{\lambda_i^k}{j!} e^{-\lambda_i}, & k = m \\ 0, & \text{otherwise} \end{cases} \tag{7.5}$$

In particular, the average number of attacks on each node $i \in \mathbb{N}$, denoted by μ_i, is

$$\mu_i = \sum_{k=1}^{m-1} k \cdot \frac{\lambda_i^k}{k!} e^{-\lambda_i} + m \sum_{k=m}^{\infty} \frac{\lambda_i^k}{k!} e^{-\lambda_i} \tag{7.6}$$

Let $\lambda = (\lambda_1, \ldots, \lambda_N)$ and assume a foresight belief on λ in terms of a prior distribution $\mathcal{F} = (\mathcal{F}_1, \ldots, \mathcal{F}_N)$ over λ, where $\lambda_i \sim \mathcal{F}_i$ for all $i \in \mathbb{N}$. Specifically, it is assumed that each $\mathcal{F}_i$ is a gamma distribution with parameters (α_i, β_i) and all $\mathcal{F}_i$'s are mutually independent with each other. Whenever k cyber-attacks are observed at node $i \in \mathbb{N}$, the parameters (α_i, β_i) are updated through Bayesian posterior as $(\alpha_i, \beta_i) \to (\alpha_i + k, \beta_i + 1)$.

In summary, the Bayesian prior method [259] is employed for the model parameters and the problem is formulated as a Bayesian adversarial MNB model.

7.2 Cost of MNB Model

In this section, a Bayesian minmax type regret function is constructed, which is subject to the learning context.

7.2.1 Regret Function of Worst Case

At time $t = 1, 2, \ldots$, a cost $c_{i,t}$ is calculated from an optimization process for all $i \in \mathbb{N}$. Without loss of generality, it is assumed that all $c_{i,t}$'s are normalized such that $c_{i,t} \in [0, 1/m]$ for all $i \in \mathbb{N}$ and $t = 1, 2, \ldots$. The process uses some inputs (e.g., external factors) that are decided by an adversary (environment).

Without ambiguity, the cost vector $c_t = (c_{1,t}, \ldots, c_{N,t})$ is assumed to be decided by the adversary at each time $t = 1, 2, \ldots$. If node $i \in \mathbb{N}$ is probed, the operator avoids incurring a total cost of $X(i, t) = K_{i,t} \cdot c_{i,t}$, $X(i, t) \in [0, 1]$. Equivalently, a reward of $K_{i,t} \cdot c_{i,t}$ is achieved by probing node π_t at time t. An admissible policy π is a sequence of mappings $\pi = \{\pi_1, \pi_2, \ldots\}$, where each π_t is a mapping from historical observations to the set of all probability distributions on $\mathbb{N}$. To distinguish, let $\{i_1, i_2, \ldots\}$ represent the sequence of chosen nodes. The axioms of the problem are formulated in the problem formulation below:

For each time $t = 1, 2, \ldots$.

- The adversary decides a cost $c_{i,t}$ for all $i \in \mathbb{N}$.
- The operator probes a node $i \in \mathbb{N}$ based on π_t and observes the number of cyber-attacks on this to be $k_{i_t,t}$.
- All $c_{i,t}$'s details are revealed to the operator.
- A reward $k_{i_t,t} \cdot c_{i_t,t}$ is achieved by the operator.

For notational convenience, let $\theta = (\alpha_1, \beta_1, \ldots, \alpha_N, \beta_N)$ denote the initial parameter vector and the cost sequence $\vec{c}_T = \{c_1, \ldots, c_T\}$ up to time T. To proceed, the optimal policy and the regret function is firstly formulated given the parameter vector θ and cost sequence $\vec{c}_T$.

The mean reward of node $i \in \mathbb{N}$ at time t is $\mathbf{E}(K_{i,t} \cdot c_{i,t}) = \mu_i \cdot c_{i,t}$. Since the reward sequence from each node is not stationary due to $c_{i,t}$, for all $t = 1, 2, \ldots$, the optimal node is defined as $i_t^* \triangleq \operatorname{argmax}_{i \in \mathbb{N}} \mu_i \cdot c_{i,t}$ and a non-stationary optimal policy is supposed to choose node i_t^* at time $t = 1, 2, \ldots$. Thus, for any admissible policy π, given θ and $\vec{c}_T$, the regret function up to time T is defined as

$$R^\pi(\lambda, \vec{c}_T, T) \triangleq \sum_{t=1}^{T} \mu_{i_t^*} \cdot c_{i_t^*,t} - \mathbf{E}^\pi \left(\sum_{t=1}^{T} \mu_{\pi_t} \cdot c_{\pi_t,t} | \lambda, \vec{c}_T \right) \tag{7.7}$$

where $\mathbf{E}^\pi$ means the expectation taken with respect to the random sequence $\{i_1, \ldots, i_T\}$ generated by π.

A constraint is on the adversary by introducing a sequence of sets $\{\mathbb{C}_T \subset [0, 1]^T\}_{T \geq 2}$, that for all time $t = 1, 2, \ldots, \vec{c}_T \in \mathbb{C}_T$. Then, given θ, the regret function with respect to the worst case up to time $T > 0$ is

$$\sup_{\vec{c}_T \in \mathbb{C}_T} R^\pi(\lambda, \vec{c}_T, T) \tag{7.8}$$

Note that $\sup_{\vec{c}_T \in \mathbb{C}_T} R^{\pi}(\lambda, \vec{c}_T, T)$ is supposed to measure the performance with the action sequence uniformly on all possible cost sequences. Finally, incorporating the prior distribution F on θ, the Bayesian regret function with respect to the worst case is redefined as

$$R^{\pi}(\mathbb{C}_T, T) \triangleq E_{\lambda \sim F}[\sup_{\vec{c}_T \in \mathbb{C}_T} R^{\pi}(\lambda, \vec{c}_T, T)] \tag{7.9}$$

To distinguish between the regret function in (7.7), (7.8) and (7.9), they are named as regret, sup regret and Bayesian sup regret, respectively, for brevity.

7.2.2 Upper Bound on Cost

By analyzing some real datasets, such as the Elia Grid, Belgium, the sequence of set $\{\mathbb{C}_T \subset [0,1]^T\}_{T \geq 2}$ is formulated with constraint on the cost. It is concluded from the statistical analysis of the real data that the temporal variations of the cost sequence of each node $i \in \mathbb{N}$, i.e., $\{|c_{i,t+1} - c_{i,t}|\}_{t \geq 1}$, are stationary and uniformly upper bounded by a value, which only relies on i and is small compared to the average value of $\{c_{i,t}\}_{t \geq 1}$.

Therefore, it is assumed that for each node the cumulative temporal variation up to time $t \geq 1$ is upper bounded by a linear function of t. Based on this assumption, a uniform upper bound on the cumulative temporal variation for all nodes is introduced and $\{\mathbb{C}_T \subset [0,1]^T\}_{T \geq 2}$ is constructed as follows:

$$\mathbb{C}_T \triangleq \left\{ \{c_{i,t}\}_{t=1}^T \in [01]^T, i \in \mathbb{N} : \sum_{t=1}^T \max_{i \in \mathbb{N}} |c_{i,t+1} - c_{i,t}| \leq \mathcal{V}_T \right\}, \forall T \geq 2 \tag{7.10}$$

where $\{\mathcal{V}_T\}_{T \geq 1}$ is a sequence of positive numbers. Since $c_{i,t} \in [0, 1/m]$, $\forall i \in \mathbb{N}$ and $t \geq 1$, $\max_{i \in \mathbb{N}} |c_{i,t+1} - c_{i,t}| \ll 1/m$ holds for all $t \geq 1$. Thus, $\mathcal{V}_T \leq O(T)$ and $\mathcal{V}_T/T \ll 1/m$. Meanwhile, $\mathcal{V}_T$ is monotonically increasing in T. Based on these considerations, the following assumption on $\mathcal{V}_T$ is proposed.

Assumption 7.1 There exist $T_0 \geq 1$, such that for all $T \geq T_0$, $1/m \leq \mathcal{V}_T \leq T/m$.

The restriction of $T \geq T_0$ for some T_0 in *Assumption* 7.1 is necessary. Actually, the condition $1/m \leq \mathcal{V}_T$ cannot be directly concluded based on the previous discussions. Since $\mathcal{V}_T$ is the accumulation of $\max_{i \in \mathbb{N}} |c_{i,t+1} - c_{i,t}|$ and we suppose that $\max_{i \in \mathbb{N}} |c_{i,t+1} - c_{i,t}| \ll 1/m$.

Hence, when T is small, $\mathcal{V}_T$ may not satisfy the condition $1/m \leq \mathcal{V}_T$. It is worth noting that justification of the existence of T_0 is able to guarantee the performance of the proposed algorithm, while an exact value of T_0 is not necessary. Prior knowledge can be incorporated to ensure the existence of T_0.

For example, if the difference term $\max_{i \in \mathbb{N}} |c_{i,t+1} - c_{i,t}|$ fails to rapidly become infinitely small (which is natural when the cost sequences are nonstationary), then $\mathcal{V}_T$ will surely be larger than $1/m$ when T is large enough. Examples of securing an estimated value for T_0 can be found in the smart grid application of Section 7.4, where $\mathcal{V}_T$ is estimated to grow linearly in T and a threshold time for $\mathcal{V}_T$ to exceed $1/m$ can be easily calculated.

7.3 Thompson-Hedge Algorithm

In this section, an online learning algorithm is developed to optimize the Bayesian sup regret for the problem. The algorithm has two layers. In the outer layer, at each time $t \geq 1$, the algorithm uses Thompson sampling (posterior sampling) method to sample a parameter vector $(\lambda_1, \ldots, \lambda_N)$ from the posterior distributions of all λ_i's based on the historical observations.

Then, in the inner layer, a so-called sub-algorithm is fed with the sampled parameters. The sub-algorithm returns an action $\pi_t \in \mathbb{N}$. At the end of this loop, the algorithm chooses node π_t, observes the reward from π_t and updates the posterior distribution of λ_{π_t}. In the end of this section, a theorem is provided to characterize the convergence rate of the regret function.

7.3.1 Hedge Algorithm

Since Hedge algorithm is applied as the sub-algorithm in the inner layer, the proposed algorithm is named as Thompson-Hedge algorithm. Before formally presenting proposed algorithm, the Hedge algorithm is firstly introduced.

Hedge algorithm is a classical algorithm designed for the adversarial MNB problem with full feedback [374]. It is a randomized algorithm that maintains a weight $\omega_t(i)$ for each $i \in \mathbb{N}$. Hedge algorithm then chooses node $i \in \mathbb{N}$ with a probability proportional to $\omega_t(i)$ at time t. Subsequently, the weight for each node is updated according to the observed reward of this node. The algorithm chooses a size Δ and restarts updating the weights of each node every Δ times.

The set of intervals between two successive restarting epochs (including the first restarting time) is called a batch. Since Hedge algorithm is designed for non-Bayesian adversarial MNB, in order to use it under the Bayesian framework, one needs to modify the original Hedge algorithm and feed it by the value of parameter λ. Denote Hedge (λ) as the modified Hedge algorithm used in the developed Bayesian adversarial MNB fed by λ. The algorithm is summarized in Algorithm 7.1.

Algorithm 7.1   Hedge (λ) Algorithm Initialization

Initialization: Parameter vector λ, $\varepsilon = (1 - \sqrt{\ln N / 2T})^{-1}$, calculate μ_i using λ_i.

```
1)   for b = 1, 2, ···, [T/Δ] do
2)       For all i ∈ N, ω₁ⁱ = 1.
3)       while t ≤  min {T, b · Δ} do
4)           For each i ∈ N, set
```
$$p_t^i = \frac{\omega_t^i}{\sum_{j \in \mathbb{N}} \omega_t^j}.$$
```
5)           Choose a node π_t according to the probability
             distribution {p_t^i}_{i∈N} and receive a reward k_{π_t,t} · c_{π_t,t}.
6)           For each node i ∈ N, update
```
$$\omega_{t+1}^i = \omega_t^i \varepsilon^{\mu_i \cdot c_{\pi_t,t}}$$
```
7)           Set t → t + 1.
8)       end while
9)   end for
```

Hedge (λ) follows the paradigm of the classical Hedge algorithm. The following lemma shows that if the value of parameter λ fed to Hedge (λ) is the true parameter of the Bayesian adversarial MNB model, then Hedge (λ) retains a convergent sup regret with a known upper bound.

Lemma 7.1 If the input λ in the Hedge algorithm is the true model parameter and the batch size is chosen to be $\Delta = \left\lceil (\ln N)^{\frac{1}{3}} (T/m\mathcal{V}_T)^{2/3} \right\rceil$, then for all $T \geq N$, the sup regret by the Hedge (λ) algorithm is upper bounded by

$$\sup_{\vec{c}_T \in \mathbb{C}_T} R^{\pi}(\lambda, \vec{c}_T, T) \leq (8 + 2\sqrt{2})(m\mathcal{V}_T \ln N)^{1/3} (T)^{2/3} \qquad (7.11)$$

Detailed proof of *Lemma* 7.1 is given [375]. In the proof of the main result of *Theorem* 7.1, the conclusion of *Lemma* 7.1 is used as an intermediate benchmark result.

In the proposed Thompson-Hedge algorithm, parameter λ is sampled from its posterior distribution in each epoch. Then, the sampled parameter is fed to the inner algorithm Hedge (λ) and a node is chosen correspondingly. In particular, the algorithm chooses a node according to the probability weight of each node.

Then, the algorithm observes the costs and updates the posterior distribution of λ. Next, a sampled parameter $\hat{\lambda}$ is drawn from the posterior distribution of λ. Finally, the algorithm updates the weight of each node for next epoch using both the observed costs and $\hat{\lambda}$.

7.3.2 Details of Thompson-Hedge Algorithm

Gamma(α, β) is used to represent Gamma distribution with parameters (α, β). The proposed Thompson-Hedge algorithm is summarized in Algorithm 7.2.

Algorithm 7.2 Thompson-Hedge Algorithm
Initialization: Parameter vector θ, batch size Δ, $\varepsilon = (1 - \sqrt{\ln N / 2T})^{-1}$, $t = 1$.

```
1)   for b = 1, 2, ···, [T/Δ] do
2)       For all i ∈ N, ωᵢ₁ = 1.
3)       while t ≤ min {T, b · Δ} do
4)           For each i ∈ N, set
```
$$p_t^i = \frac{\omega_t^i}{\sum_{j \in N} \omega_t^j}.$$
```
5)           Choose a node iₜ according to the probability
             distribution {pₜⁱ}ᵢ∈ℕ and receive a reward k_{iₜ,t} · c_{iₜ,t}.
6)           For node iₜ, update (α_{iₜ}, β_{iₜ}) → (α_{iₜ} + k_{iₜ,t}, β_{iₜ} + 1).
7)           For each node i ∈ N, sample λ̂ ~ Gamma(αᵢ, βᵢ),
             calculate μ̂ᵢ using λ̂ᵢ, and update
```
$$\omega_{t+1}^i = \omega_t^i \varepsilon^{\hat{\mu}_i \cdot c_{i_t,t}}$$
```
8)           Set t → t + 1.
9)       end while
10)  end for
```

Before the main result (Theorem 7.1) is presented, the following lemma is introduced that will be used for the proof of the main result.

Lemma 7.2 [[376], Lemma 2] If F is the true distribution of λ, and $\hat{\lambda}$ is the sampled parameter in epoch t, then for any $\sigma(\mathbb{H}_t)$ - measurable function g, it follows

$$\mathbf{E}_\lambda[g(\lambda)|\mathbb{H}_t] = \mathbf{E}_{\hat{\lambda}}[g(\hat{\lambda})|\mathbb{H}_t]. \tag{7.12}$$

Lemma 7.2 shows a central result in Bayesian learning area. The sampled parameter from the posterior distribution each time can be "considered" as the true parameter in the sense that any deterministic function using it as an argument independently has the same expectation. On the basis of *Lemma 7.1* and *Lemma 7.2*, the main theorem is presented as below, which establishes an upper bound on the Bayesian sup regret for the proposed Thompson Hedge algorithm and thus indicates its convergence rate.

Theorem 7.1 Let π be the Thompson-Hedge algorithm with batch size $\Delta = (mV_T \ln N)^{1/3}$ $(T)^{2/3}$, $\mathbb{C}_T$ be defined in (7.10), and T_0 be the same value as defined in Assumption 7.1, then for all $T \geq T_0$, the Bayesian sup regret $R^\pi(\mathbb{C}_T, T)$ can be upper bounded by

$$R^\pi(\mathbb{C}_T, T) \leq (8 + 2\sqrt{2})(mV_T \ln N)^{1/3}(T)^{2/3} \tag{7.13}$$

Proof: The basic idea for the proof is sketched as follows.

The Bayesian sup regret by Thompson-Hedge algorithm is firstly decomposed into two parts: the regret of the introduced Hedge (λ) algorithm and the difference of performance between Thompson-Hedge and Hedge algorithms. Since the sup regret by Hedge algorithm is bounded by *Lemma 7.1*, *Theorem 7.1* can then be proved by upper bounding the performance difference between the two algorithms.

To distinguish between two algorithms in the work, π^{TH} and π^H are used for Thompson-Hedge and Hedge algorithms correspondingly. Moreover R^{TH} and R^H are denoted as the regret function of π^{TH} and π^H, respectively.

7.3.2.1 Separation of Target Regret
In Section 7.3.2.1, the Bayesian sup regret R^{TH} is separated into a combination of two terms.

In Section 7.3.2.2 and Section 7.3.2.3, the upper bound of these two terms are obtained separately and therefore, the upper bound of R^{TH} can be readily obtained. The separation of R^{TH} is given as below.

$$R^{TH}(\mathbb{C}_T, T) = \mathbf{E}_{\lambda \sim F}\left[\sup_{\vec{c}_T \in \mathbb{C}_T}\left\{\sum_{t=1}^T \mu_{i_t^*} \cdot c_{i_t^*,t} - \mathbf{E}^{TH}\left(\sum_{t=1}^T \mu_{i_t} \cdot c_{i_t,t}|\lambda, \vec{c}_T\right)\right\}\right]$$

$$\leq \mathbf{E}_{\lambda \sim F}\left[\sup_{\vec{c}_T \in \mathbb{C}_T}\left\{\sum_{t=1}^T \mu_{i_t^*} \cdot c_{i_t^*,t} - \mathbf{E}^H\left(\sum_{t=1}^T \mu_{i_t} \cdot c_{i_t,t}|\lambda, \vec{c}_T\right)\right\}\right] \tag{7.14}$$

$$+ \mathbf{E}_{\lambda \sim F}\left[\sup_{\vec{c}_T \in \mathbb{C}_T}\left\{\mathbf{E}^H\left(\sum_{t=1}^T \mu_{i_t} \cdot c_{i_t,t}|\lambda, \vec{c}_T\right) - \mathbf{E}^{TH}\left(\sum_{t=1}^T \mu_{i_t} \cdot c_{i_t,t}|\lambda, \vec{c}_T\right)\right\}\right]$$

For notational convenience, denote the above two terms in the two square brackets by Λ_1 and Λ_2, respectively. The relation in (7.14) can be rewritten as

$$R^{TH} = \Lambda_1 + \Lambda_2 \tag{7.15}$$

Note that Λ_1 is closely associated with the sup regret of π^H and can be bounded based on the result in *Lemma* 7.1. Λ_2 is the difference between the return of π^{TH} and π^H that will be bounded in the following section.

7.3.2.2 Upper Bound of Λ_1

Λ_1 is rewritten as

$$\Lambda_1 = \mathbf{E}_{\lambda \sim \mathcal{F}}[\sup_{\vec{c}_T \in \mathbb{C}_T} R^H(\lambda, \vec{c}_T, T)] \tag{7.16}$$

Note that the upper bound in *Lemma* 7.1 holds for any true model parameter λ. Thus the upper bound still holds after taking expectation on λ on both sides of (7.11). Therefore, it holds that

$$\Lambda_1 \leq (8 + 2\sqrt{2})(m\mathcal{V}_T \ln N)^{1/3}(T)^{2/3} \tag{7.17}$$

7.3.2.3 Upper Bound of Λ_2

The following clarification is made for notational convenience. λ_t is used for the sampled parameter by Thompson-Hedge algorithm at time t and λ for the true model parameter which is the input of Hedge algorithm. To bound the Bayesian sup regret by Thompson Hedge algorithm, the difference between the Bayesian sup regret functions by Thompson-Hedge and Hedge algorithms, respectively, is firstly bounded, namely,

$$\Lambda_2 = \mathbf{E}_{\lambda \sim \mathcal{F}}\left[\sup_{\vec{c}_T \in \mathbb{C}_T}\left\{\mathbf{E}^{TH}\left(\sum_{t=1}^{T}\mu_{i_t} \cdot c_{i_t,t}|\lambda, \vec{c}_T\right) - \mathbf{E}^{H}\left(\sum_{t=1}^{T}\mu_{i_t} \cdot c_{i_t,t}|\lambda, \vec{c}_T\right)\right\}\right] \tag{7.18}$$

where $\mathbf{E}^{TH}$ and $\mathbf{E}^{H}$ have the same meaning as $\mathbf{E}^\pi$ given π, for Thompson-Hedge algorithm and Hedge algorithm, respectively. Denote the observation history before time $t \geq 2$ as

$$\mathbb{H}_t = \{i_1, k_{i_1,1}, c_1, \ldots, i_{t-1}, k_{i_{t-1},t-1}, c_{t-1}\}.$$

Conditioned on the observation history, the probability weight function $p_t(\cdot)$ in both algorithms is functions of λ and $\lambda(t)$ correspondingly. Moreover, let $p_t(\cdot)$ and $\widetilde{p}_t(\cdot)$ be the probability functions of Thompson-Hedge and Hedge algorithms, respectively.

For any fixed $\vec{c}_T \in \mathbb{C}_T$, the following relation holds,

$$\mathbf{E}^{TH}\left(\sum_{t=1}^{T}\mu_{i_t} \cdot c_{i_t,t}|\lambda, \vec{c}_T\right) - \mathbf{E}^{H}\left(\sum_{t=1}^{T}\mu_{i_t} \cdot c_{i_t,t}|\lambda, \vec{c}_T\right)$$

$$= \mathbf{E}^{TH}\left(\sum_{t=1}^{T}\mu_{i_t} \cdot c_{i_t,t}|\lambda, \vec{c}_T, \mathbb{H}_T\right) - \mathbf{E}^{H}\left(\sum_{t=1}^{T}\mu_{i_t} \cdot c_{i_t,t}|\lambda, \vec{c}_T, \mathbb{H}_T\right) \tag{7.19}$$

$$= \sum_{t=1}^{T}\mathbf{E}_{\lambda_t}[p_t^i(\lambda_t) - \widetilde{p}_t^i(\lambda)|\mathbb{H}_t]\mu_i \cdot c_{i,t}$$

At any time $t \geq 1$, note that λ_t is the sampled parameter from the same posterior distribution as the true λ. Meanwhile, $p_t^i(\ )$ and $\widetilde{p}_t^i(\cdot)$ are the same deterministic function based on $\mathbb{H}_t$. According to *Lemma* 7.2, it follows

$$\mathbf{E}_{\lambda_t,\lambda}[p_t^i(\lambda_t) - \widetilde{p}_t^i(\lambda)|\mathbb{H}_t]\mu_i \cdot c_{i,t} = 0 \tag{7.20}$$

Therefore, for any fixed $\vec{c}_T \in \mathbb{C}_T$, it holds that

$$\mathbf{E}_{\lambda \sim \mathcal{F}}\left[\sup_{\vec{c}_T \in \mathbb{C}_T}\left\{\mathbf{E}^{TH}\left(\sum_{t=1}^{T}\mu_{i_t}\cdot c_{i,t}|\lambda,\vec{c}_T\right) - \mathbf{E}^{H}\left(\sum_{t=1}^{T}\mu_{i_t}\cdot c_{i,t}|\lambda,\vec{c}_T\right)\right\}\right]$$

$$= \mathbf{E}\left(\mathbf{E}_{\lambda \sim \mathcal{F}}\left[\left\{\mathbf{E}^{TH}\left(\sum_{t=1}^{T}\mu_{i_t}\cdot c_{i,t}|\lambda,\vec{c}_T,\mathbb{H}_T\right) - \mathbf{E}^{H}\left(\sum_{t=1}^{T}\mu_{i_t}\cdot c_{i,t}|\lambda,\vec{c}_T,\mathbb{H}_T\right)\right\}\right]\right)$$

(7.21)

$$= \mathbf{E}(\mathbf{E}_{\lambda_t,\lambda}[p_t^i(\lambda_t) - \widetilde{p}_t^i(\lambda)|\mathbb{H}_t]\mu_i\cdot c_{i,t}) = 0$$

Note that the relation above holds for any $\vec{c}_T \in \mathbb{C}_T$, which leads to

$$\mathbf{E}_{\lambda \sim \mathcal{F}}\left[\sup_{\vec{c}_T \in \mathbb{C}_T}\left\{\mathbf{E}^{TH}\left(\sum_{t=1}^{T}\mu_{i_t}\cdot c_{i,t}|\lambda,\vec{c}_T\right) - \mathbf{E}^{H}\left(\sum_{t=1}^{T}\mu_{i_t}\cdot c_{i,t}|\lambda,\vec{c}_T\right)\right\}\right] = 0 \quad (7.22)$$

7.3.2.4 Upper Bound of Regret R^{TH}

Finally, the upper bound of the Bayesian sup regret of Thompson-Hedge algorithm is obtained by combining the results in Section 7.3.2.1 to Section 7.3.2.3.

In particular, it holds that

$$R^{TH}(\mathbb{C}_T, T) \le \Lambda_1 + \Lambda_2 = \Lambda_1 + 0$$

$$\le (8 + 2\sqrt{2})(m\mathcal{V}_T \ln N)^{1/3}(T)^{2/3} \quad (7.23)$$

Therefore, the proof is concluded.

Remark 7.1 Relevant research that considers a similar problem can be found in [377]. In [377], the classical EXP3 type algorithm was used and an upper bound of the order $O((m\mathcal{V}_T \ln N)^{1/3}(T)^{2/3})$ was obtained for the sup regret.

Since the upper bound holds uniformly on the parameter space, the same upper bound also holds for the Bayesian sup regret. If the constant $(8 + 2\sqrt{2})$ in (7.22) is neglected, the bound outperforms the bound in [377] by a term of $O(N^{1/3})$, which implies that the performance of the proposed Thompson-Hedge algorithm is less sensitive to the number of nodes N.

It indicates that when considering problems with large scales, the proposed algorithm is supposed to retain a Bayesian sup regret that converges relatively faster. Meanwhile, [377] constructs a lower bound of the order $O(\mathcal{V}_T^{1/3}T^{2/3})$ on the regret by any algorithm. Similarly, the lower bound also holds uniformly on all the parameters and adapts to the problem. Therefore, the lower bound shows that the proposed algorithm achieves the order optimality.

7.4 Applications to Smart Grids

The model is formulated following the learning context in Section 7.2 and an online learning algorithm is developed in Section 7.3. This section presents an application case to demonstrate the applicability of the proposed model and method in practical smart grids.

In particular, *Assumption* 7.1 plays a central role in the proposed model. A real data set is used to verify that *Assumption* 7.1 may hold in reality, and thus the proposed method is practical. Section 7.4.1 introduces the procedure for calculating the reward $c_{i,t}$ in the smart grid. In Section 7.4.2, linear regression method is used to show a linear growth rate of the critical quantity V_T in *Assumption* 7.1, and thus it is concluded that *Assumption* 7.1 holds for the selected data set.

7.4.1 Operation Cost of Smart Grids

To facilitate reading, meanings of the variables used in calculating the operation cost of the smart grid, are displayed in different categories as shown in Section 4.4.3. The linear DC-OPF model is introduced to interpret the physical meaning of $c_{i,t}$ in the smart grid.

At time t, the operation state of the Distributed Generation Systems (DGS) is denoted by following vector:

$$\vartheta_t = [P_{i,j}^t, L_i^t], \tag{7.24}$$

where the power sources considered in this work consists of natural gas plant, biomass plant, wind farm, Photovoltaic Power (PV) farm and Energy Storage System (ESS). These data are all sampled from the historical dataset provided by the Elia Grid, Belgium.

To calculate $c_{i,t}$, the first objective is to achieve the minimal cost, denoted by Co^{ϑ_t}, in the presence of load shedding without conflicting the physical characters of power systems, by solving the following linear optimization problem.

$$\text{Min}Co^{\vartheta_t} = \sum_{i \in N} \sum_{j \in \mathbb{P}} (Cs_j - Ep^{\vartheta_t}) P_{i,j}^t$$

$$+ \sum_{(i,i') \in \mathbb{F}} Cf_{i,i'} |B_{i,i'}(\theta_i - \theta_{i'})| + (Cp + Ep^{\vartheta_t}) \sum_{i=1}^{N} \overline{L}_i^t \tag{7.25}$$

subject to

$$L_i^t - \overline{L}_i^t - \sum_{j \in \mathbb{P}} P_{i,j}^t - \sum_{i=1}^{N} B_{i,i'}(\theta_i - \theta_{i'}) = 0, \tag{7.26}$$

$$0 \le P_{i,j}^t \le \overline{P}_{i,j}, j \neq ESS, \tag{7.27}$$

$$|P_{i,j}^t \cdot T_s| \le \min\{(1 - \varphi) \cdot \overline{C}_{ESS}, \overline{P}_{i,j} \cdot T_s\}, j = ESS, \tag{7.28}$$

$$|B_{i,i'}(\theta_i - \theta_{i'})| \le (1 - s_{i,i'})V \cdot A_{i,i'}, \tag{7.29}$$

where constraint (7.26) requires that the power generated and consumed is balancing at each node of DGSs, constraint (7.27) requires that the power generated should not be larger than the rated power, constraint (7.28) indicates that the charging or discharging of ESS should not be larger than the remaining capacity or nominal rate, and constraint (7.29) indicates that the power flow between two nodes should not be larger than the capacity of the transmission line.

This linear Direct Current (DC) optimal power flow model is configured in the Matlab and can be solved using the well-known Simplex method, where the values of configuration parameters have been given in [366] and the computation time of operation cost for each practical operation state is around 0.006 second in Gurobi. This computation time contributes most to the total simulation time and therefore the proposed algorithm can be implemented in real time.

The operational variables are that ϑ_t is the operation state of the smart grid at time t, $P^t_{i,j}$ [kW] is the output of type j power source at Node i, $\overline{P}_{i,j}$ is the rate power of type j power source at Node i, L^t_i is the power load at Node i, $\overline{L}^t_i$ is the load shedding, i.e., power demand not supplied, at Node i, φ is the state of charge of the energy storage system and $\overline{C}_{ESS}$ is the power capacity of the energy storage system.

The configuration parameters of the smart grid are that $B_{i,i'}$ [1/Ω] is susceptance of the pairs of Nodes (i, i'), θ_i is the voltage angle at Node i, $\Delta\theta$ is the voltage angle difference $(\theta_i - \theta_{i'})$ of two nearby nodes, $A_{i,i'}$ [A] is the ampacity of the pairs of Nodes (i, i') and V [kW] is the nominal voltage of the smart grid.

The cost and price coefficients are that Co^{ϑ_t} is the operation cost of the smart grid subject to operation state ϑ_t, Cs_j is the variable operation cost for power source j, $Cf_{i,i'}$ is the variable operation cost for feeder (i, i'), Cp is the penalty cost for power demand not supplied, and Ep^{ϑ_t} is the energy price associated with operation state ϑ_t. T_s is the duration of ϑ_t, $\mathbb{P}$ denotes a subset of power sources, $\mathbb{N}$ denotes the set of all nodes, and $\mathbb{F}$ denotes the set of node pairs with transmission line between them. $s_{i,i'} = 1$ indicates that power cannot be transmitted between Node i and Node i' due to the successful cyber-attack on Node i.

The physical meaning of $c_{i,t}$ in the smart grid can be explained as the difference between the operation cost of the DGS without cyber-attacks and the operation cost of the DGS given the Node $i^*$is temporarily unavailable caused by a successful cyber-attack. Therefore, if we probe Node i at time t, the reward function $c_{i,t}$ is defined as

$$c_{i,t} \triangleq |Co^{\vartheta_t}(P^t_{i,j}, L^t_i, \Delta\theta, s_{i,i'} = 0) - Co^{\vartheta_t}(P^t_{i,j}, L^t_i, \Delta\theta, s_{i,i'} = 1)|, \tag{7.30}$$

where $Co^{\vartheta_t}(P^t_{i,j}, L^t_i, \Delta\theta, s_{i,i'} = 0)$ is the operation cost of the DGS without cyber-attacks, and $Co^{\vartheta_t}(P^t_{i,j}, L^t_i, \Delta\theta, s_{i,i'} = 1)$ is the operation cost of the DGS given Node i is unavailable caused by the successful cyber-attack on Node i.

That is to say, the reward $c_{i,t}$ can be calculated via solving the linear optimization problem, defined by (24) to (27), twice (with/without cyber-attacks), where the input ϑ_t is drawn from the dataset of Elia Grid, Belgium.

7.4.2 Numerical Analysis of Cost Sequences

In this section, numerical analysis is presented based on a real data set to verify that *Assumption 7.1* holds in reality. Note that *Assumption 7.1* implies a linear or sub-linear upper bound in terms of T on $\mathcal{V}_T$.

Therefore, if $\mathcal{V}_T$ has a linear or sub-linear growth rate in T, then *Assumption 7.1* is supposed to hold by choosing a proper value for m. Thus, linear regression is performed on the sequence of $\mathcal{V}_T$ against time T. The realistic grid data from the Elia is used, which provides data from the Belgian electricity market system. The underlying electricity network (a subgrid of Elia Grid) is shown in Figure 7.1.

In the DGS, the dataset of Elia Grid, Belgium is recorded every 15 minutes, which means that the Optimal Power Flow (OPF) model will be performed and generate one attack cost $c_{i,t}$ every 15 minutes. The DGS under cyber-attacks is investigated over one week, which indicates that a specific dataset with 672 successive observations of the attack cost is selected

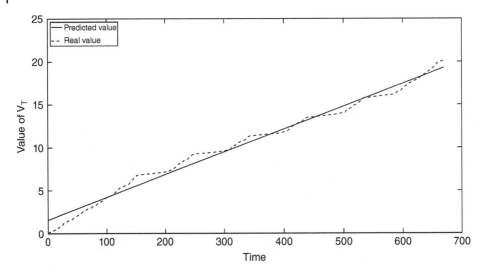

Figure 7.3 Data scatter plot and linear prediction of $\mathcal{V}_T$.

Table 7.1 Regression results.

	Coefficients	Standard error	t Stat	P-value
Intercept	1.560	0.04711	32.04	<0.01
T	0.02648	0.0001212	218.5	0

Table 7.2 ANOVA results.

	df	SS	MS	F	Significance
Regression	1	17655	17655	47734	≤0.01
Residual	669	247.4	0.3699		
Total	670	17902.4			

for illustration. m is chosen to be 4 and the data has been normalized such that $c_{i,t} \in [0, 1/4]$. The scatter plot is presented in Figure 7.3.

From Figure 7.3, a significant linear relation between $\mathcal{V}_T$ and T can be observed. For further justification, the numerical results of the linear regression are presented in Table 7.1.

According to the results in Table 7.1, the coefficient for T is about 0.03, which means that *Assumption* 7.1 holds with T_0 chosen to be 9. As a significance test for the regression model, results of Analysis of variance (ANOVA) for the regression model is shown in Table 7.2.

From the results in Table 7.2, the significance level is below 0.01, which means that the linear relation between $\mathcal{V}_T$ and T for this data set is significant and the model is justified.

7.5 Performance of Thompson-Hedge Algorithm

In this section, the performance of the proposed Thompson-Hedge algorithm is illustrated through simulation studies. In Section 7.5.1, a comparative study is conducted between the performances of the proposed Thompson-Hedge algorithm and the R.EXP3 algorithm which was recently proposed in [377].

Subsequently, in Section 7.5.2, sensitivity analysis is conducted to investigate the influence of constraint V_T on the cost sequence.

7.5.1 Comparison Study Against R.EXP3

This section compares the performance of the proposed Thompson-Hedge algorithm with the R.EXP3 algorithm that was also designed for the same adversarial problem with constrained variation on cost sequence.

Both algorithms use batch methods from the original EXP3 algorithm, i.e., the time horizon is divided into small batches and the algorithm "restarts" at the beginning of each batch. Meanwhile, the R.EXP3 algorithm is also a randomized algorithm, retains a weight function for each node and updates the weights each time.

Different from the proposed Thompson-Hedge algorithm, in the R.EXP3 algorithm, if node $i \in \mathbb{N}$ is chosen at any time, the total reward $k_{i,t} \cdot c_{i,t}$ is considered as a whole adversarial reward. Since the attack number $k_{i',t}$ is unknown if $i \neq i'$, the R.EXP3 is supposed to ignore $c_{i',t}$ and treat the problem as a typical MNB model with bandit feedback.

The simulation is executed under $N = 10$ and $N = 20$. To initialize, the parameter $\theta = (\alpha_1, \beta_1, ..., \alpha_N, \beta_N)$ is set for all $i \in \mathbb{N}$, and two large numbers for Q and L. The preset parameter θ is used to generate $\lambda = (\lambda_1, ..., \lambda_N)$ for Q times in total. Under each generated λ, λ is used to generate the sequence $\{K_{i,t}\}_{t=1}^{T}$, and the adversarial cost sequence $\{c_{i,t}\}_{t=1}^{T}$ is artificially generated. Then, both algorithms are run based on $\{K_{i,t}\}_{t=1}^{T}$ and $\{c_{i,t}\}_{t=1}^{T}$, $\forall i \in \mathbb{N}$.

Finally, the Bayesian sup regret in the two algorithms are calculated and compared. The simulation process is summarized using pseudo codes in Algorithm 7.3. Experience from the previous work [259] and the real database from the used numerical example are integrated to determine the parameter vector θ.

Algorithm 7.3   Simulation for comparison

Initialization: Present parameter $\theta = (\alpha_1, \beta_1, ..., \alpha_N, \beta_N)$; numbers of simulation trails Q and L; time horizon T.

```
1)   for q = 1 : Q do
2)       For all i ∈ ℕ, generate λ_i from Gamma (α_i, β_i).
3)       For l = 1 : L do
4)           For t = 1 : T do
5)               For i ∈ ℕ do
6)                   Generate K_{i, t} from Poisson (λ_i).
7)                   Generate c_{i, t} artificially.
8)               end for
9)           end for
10)          Run Thompson-Hedge algorithm and R.EXP3 algorithm
             independently based on the sequences {K_{i,t}}_{t=1}^{T} and
             {c_{i,t}}_{t=1}^{T}, ∀i ∈ ℕ.
11)          Calculate the regret function R_l^{TH}(λ, c⃗_T, T) for Thompson-
             Hedge algorithm and R_l^{R3}(λ, c⃗_T, T) for R.EXP3 algorithm.
12)      end for
```

13)   Calculate the sup regret $\widetilde{R}_q^{TH} = \max_{l=1:L} R_l^{TH}$ for Thompson-Hedgealgorithm and $\widetilde{R}_q^{R3} = \max_{l=1:L} R_l^{R3}$ for R.EXP3 algorithm.

14)   **end for**

15)   Calculate the Bayesian sup regret

$$R^{TH}(\mathbb{C}_T, T) = 1/Q \sum_{q=1}^{Q} \widetilde{R}_q^{TH}$$

for Thompson-Hedge algorithm and,

$$R^{R3}(\mathbb{C}_T, T) = 1/Q \sum_{q=1}^{Q} \widetilde{R}_q^{R3}$$

for R.EXP3 algorithm.

In [259], the prior distribution is set to be Gamma $(2,2)$ in this numerical example. It is estimated that $\lambda \approx 0.25$, which may correspond to the prior distribution Gamma $(1,4)$. Therefore, two values for θ are selected to implement the simulation $\theta_1 : (\alpha_1, \beta_1) = \cdots = (\alpha_N, \beta_N) = (2,2)$ and $\theta_2 : (\alpha_1, \beta_1) = \cdots = (\alpha_N, \beta_N) = (1,4)$. According to the choices of θ, the truncation parameter m is set as 3, which ensures that the probability $\Pr(K_{i,t} > m)$ is small.

Note that in the simulation, the empirical estimation of the sup regret and Bayesian sup regret functions is used in the two algorithms. Therefore, it is necessary to set the values of Q and L large enough to make the estimation with good precision. In the simulation, it is set as $Q = L = 10^5$. To simulate the real situations, the procedures of sketching the cost sequence are given as below:

- Step 1: For all $i \in \mathbb{N}$, generate $c_{i,1}$ uniformly on $(0, 1/m)$ independently.
- Step 2: For $t = 2, \cdots, T$, generate $c_{i,t}$ independently from the uniform distribution on the overlapping interval between $(c_{i,t-1} - 1/100m, c_{i,t-1} + 1/100m, c)$ and $(0, 1/m)$.

By Step 1, simulate the random initial value of each cost sequence. Then, by Step 2, set $|c_{i,t} - c_{i,t-1}| \le 1/50m$. It can be verified that any cost sequence generated by Step 1 and Step 2 satisfies *Assumption* 7.1 with $T_0 = 50$. The simulation results under $(\alpha_1, \beta_1) = \cdots = (\alpha_N, \beta_N) = (2,2)$ are given in Figure 7.4 and Figure 7.5.

Using the same process, simulation is implemented for $(\alpha_1, \beta_1) = \cdots = (\alpha_N, \beta_N) = (1,4)$. The results are given in Figure 7.6 and Figure 7.7.

From Figure 7.4 to Figure 7.7, the proposed Thompson-Hedge algorithm outperforms the existing R.EXP3 algorithm in terms of Bayesian sup regret. In particular, Thompson-Hedge algorithm is less sensitive to the problem scale N by comparing the regret curves under $N = 10$ and $N = 20$, which is consistent with the discussion in *Remark* 7.1.

The proposed algorithm has advantages over a typical algorithm designed for MNB with bandit feedback, such as the EXP3 or R.EXP3 algorithm. The usual convergence order of the regret function by a typical algorithm is $O(N \log N)^{1/3}$ for MNB with bandit feedback in terms of the problem scale N. The proposed Thompson-Hedge algorithm feeds a set of sampled parameters to Hedge algorithm.

In the proof of *Theorem* 7.1, it is shown that using the sampled parameters is "as good as" using the true parameters under the Bayesian framework. Moreover, if the true parameters

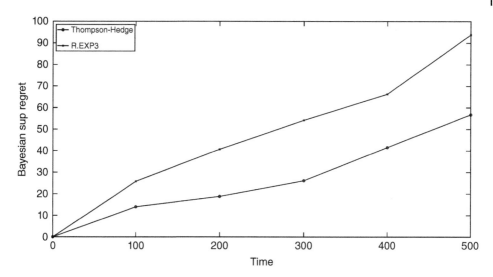

Figure 7.4 Comparison of regrets ($\theta = \theta_1, N = 10$).

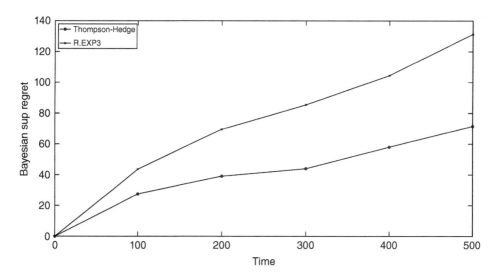

Figure 7.5 Comparison of regrets ($\theta = \theta_1, N = 20$) independently.

are known, then Hedge algorithm can solve the problem, which has a convergence order of $O(\log N)$. Therefore, the proposed Thompson-Hedge algorithm is supposed to retain a convergence rate of $O(\log N)$ for the Bayesian sup regret in the special case.

Meanwhile, upper bound in *Theorem* 7.1 is also drawn in Figure 7.4 to Figure 7.7. It is worth noting that the upper bound in *Theorem* 7.1 holds uniformly on all possible values of θ. Hence, it is concluded that when θ varies, the deviation of the regret from the upper bound also varies. In particular, as shown in the figures, the difference between the upper bound and the regret is comparatively large under $\theta = \theta_1$, $N = 20$ while small or moderate under other three parameters. However, it can be observed from the figures that the upper

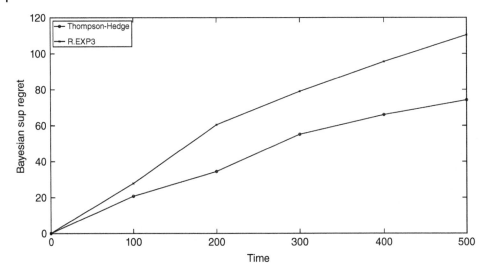

Figure 7.6 Comparison of regrets ($\theta = \theta_2, N = 10$).

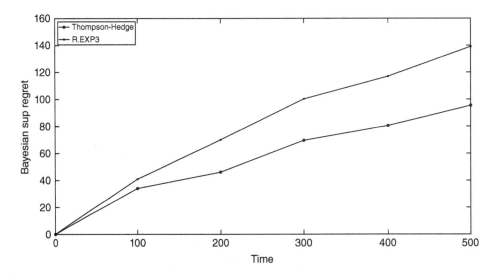

Figure 7.7 Comparison of regrets ($\theta = \theta_2, N = 20$).

bound is obviously sharper than that for the R.EXP3 algorithm in the sense that the upper bound is below the regret under R.EXP3 at some time points in all the four figures.

7.5.2 Sensitivity to the Variation

This section presents the sensitivity analysis of the proposed Thompson-Hedge algorithm to the variation constraint $\mathcal{V}_T$. In reality, according to the specific environment and workload under which the network functions, the adversary may generate the cost sequence subject to various rules. In the proposed model, the variation constraint $\mathcal{V}_T$ is used to characterize the cost sequence.

Thus, numerical results for the proposed Thompson-Hedge algorithm are presented under four levels of variation. In particular, the same steps given in Section 7.5.1 are employed, but 4 levels are selected for the variation scale in Step 2: $1/20m$, $1/50m$, $1/100m$ and $1/200m$. Other parameters are set to be identical as those in Section 7.5.1. The simulation results are illustrated under both θ_1 and θ_2 as shown in Figure 7.8 and Figure 7.9.

The results in Figure 7.8 and Figure 7.9 illustrate the different performances of the proposed Thompson-Hedge algorithm under different variation scales. The Bayesian regret monotonically increases with the variation. This is because larger variation leads

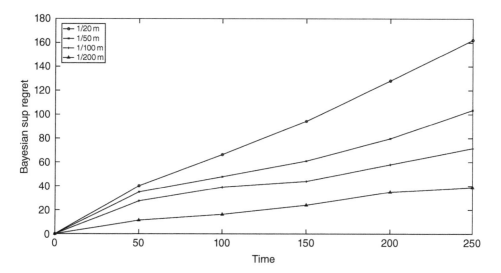

Figure 7.8 Regrets under different variations ($\theta = \theta_1, N = 20$).

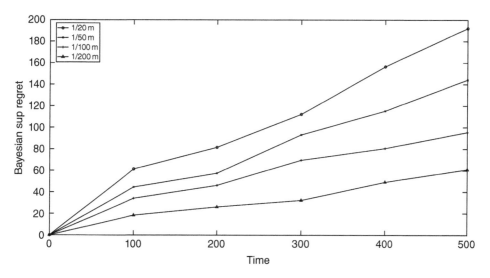

Figure 7.9 Regrets under different variations ($\theta = \theta_2, N = 20$).

to more uncertainties of the cost sequence and makes the experience from historical observations unreliable. Thus, as an adaptive online learning algorithm, the Bayesian sup regret in the proposed Thompson Hedge algorithm may increase and the performance may have fluctuations.

However, it should be noted that the incremental difference between the regret curves under $1/100m$ and $1/200m$ are not so significant compared to that between the curves under $1/50m$ and $1/20m$. This implies that when the variation is within $V_T \leq T/50m$, the regret function is less sensitive to the variation than the situation that V_T is around $1/10m$. In the numerical example, the real data set (from Elia Grid, Belgium) is examined and it appears that the corresponding V_T is around $T/100m$. Therefore, it can be concluded that the sensitivity of the proposed Thompson-Hedge algorithm to the variation may not be high in practice.

8

Recent Advances in CPS Modeling, Stability and Reliability

This chapter briefly discusses some recent studies focusing on the Cyber-Physical System (CPS) modeling, stability analysis and reliability improvement.

8.1 Modeling Techniques for CPS Components

This book mainly applies the Wiener process to describe the path of the aging CPS component, which can indicate the degradation level at arbitrary time. However, the failure time of the aging component with respect to the failure threshold and its probability distribution are not discussed but they are very important in determining the Remaining Useful Lifetime (RUL) of the CPS.

8.1.1 Inverse Gaussian Process

Various stochastic processes, e.g., Wiener process, Gamma process, inverse Gaussian processes, have been used to model the degradation of industrial systems at both component-level and system-level. For the performance degradation of valves, a linear regression model is used to fit the evolution of percentage deviation in flow rates [378]. A gamma process is used to model the wear process of spool valve in [379]. It is argued in [380] that the valve is subject to both continuous and discrete degradation processes.

They model the continuous degradation processes by a deterministic linear function and the discrete degradation processes by a compound Poisson process. In fact, the gamma process can be viewed as the limit of a compound Poisson process whose rate goes to infinity while the jump size converges to zero in certain rate [381].

Similarly, the inverse Gaussian process is also a limiting compound Poisson process and it is even more flexible than the gamma process [382] as various monotonic trends, random effects and covariates can be incorporated into an inverse Gaussian process in a handy fashion. Therefore, the inverse Gaussian process can be used as an efficient surrogate to the integrated model presented in [380] for valve degradation.

It is assumed that $Y(t)$ follows an inverse Gaussian process with the monotonically increasing mean function $\Lambda(t)$ and a dispersion parameter $\eta > 0$. $Y(t)$ has independent increments and the increment $Y(t) - Y(s)$ follows an inverse Gaussian distribution $\mathcal{IG}(\Lambda(t) - \Lambda(s), \eta[\Lambda(t) - \Lambda(s)]^2)$, $\forall t > s \geq 0$. Here, $\mathcal{IG}(\mu, \lambda)$, $\mu, \lambda > 0$, denotes the inverse

Cyber-Physical Distributed Systems: Modeling, Reliability Analysis and Applications, First Edition.
Huadong Mo, Giovanni Sansavini and Min Xie.
© 2021 John Wiley & Sons Ltd. Published 2021 by John Wiley & Sons Ltd.

Gaussian distribution with probability density [383] as

$$f(x|\mu, \lambda) = \sqrt{\frac{\lambda}{2\pi x^3}} \cdot \exp\left[-\frac{\lambda(x-\mu)^2}{2\mu^2 x}\right], \quad x > 0 \tag{8.1}$$

and cumulative distribution function (CDF)

$$F(x|\mu, \lambda) = \Phi\left[\sqrt{\frac{\lambda}{x}}\left(\frac{x}{\mu} - 1\right)\right] + \exp\left(\frac{2\lambda}{\mu}\right)\Phi\left[-\sqrt{\frac{\lambda}{x}}\left(\frac{x}{\mu} + 1\right)\right] \tag{8.2}$$

where $\Phi(\cdot)$ is the standard normal CDF.

8.1.2 Hitting Time to a Curved Boundary

In many industrial CPSs, the failure time τ of an aging unit can be defined as the time at which the degradation process first reaches a constant threshold h. The distribution of the first hitting time τ_h plays a central role in predicting remaining useful life and failure probability.

Consider the inverse Gaussian process defined in the previous section, given a constant maximum allowable degradation level of the valve h, the distribution of the hitting time t_h is given by

$$\Pr(\tau \le t) = \Pr(Y(t) \ge h)$$

$$= \Phi\left[\sqrt{\frac{\eta}{h}}(\Lambda(t) - h)\right] - \exp(2\eta\Lambda(t))\Phi\left[-\sqrt{\frac{\eta}{h}}(\Lambda(t) + h)\right]. \tag{8.3}$$

However, it is quite challenging and even infeasible to derive the exact distribution for the first hitting time of the proposed valve degradation model since the degradation threshold is not a constant.

The non-constant degradation threshold can be denoted by a function of time $h(t)$. Though it is possible to derive the exact hitting time distribution given piecewise linear hitting boundary by following the conditional argument in [384], $h(t)$ can be rather different for different control systems of CPSs under changing operational environment.

Therefore, it is necessary to find the first hitting time distribution of an inverse Gaussian process with an arbitrary curved hitting boundary. To address this challenge, we adopt the classical idea of using discrete time approximation to the continuous sample path to simulate the first hitting time τ of an inverse Gaussian process with a curved boundary $h(t)$. Particularly, it is assumed that $h(t)$ is a bounded non-negative function for any $t \ge 0$ which ensures the hitting time $\tau < +\infty$ almost surely.

The pseudocodes of the simulation algorithm are detailed in Algorithm 8.1.

Algorithm 8.1 Simulating First Hitting Time of an Inverse Gaussian Process with a Curved Boundary

- Input: mean function - $\Lambda(t)$, dispersion parameter - $\eta > 0$, hitting boundary - $h(t)$, grid step size $\Delta t > 0$, Monte Carlo Simulation (MCS) Sample Size - N, and maximum time limit - $T > 0$.
- Initialization: Compute the number of grid points $K = T/\Delta t$, generate the grid points $t_k = k\Delta t$, $k = 0, 1, 2, \ldots, K$, and compute the mean increments over the grid $\delta_k = \Lambda(t_k) - \Lambda(t_{k-1})$, $k = 1, 2, \ldots, K$.

- Run: for $i = 1, 2, \ldots, N$ do
    Step 1 Simulate the independent increments ΔY_k from $\mathcal{IG}(\delta_k, \eta\delta_k^2)$,
    Step 2 Compute the sample path $Y(t_k) = \Sigma_{i=1}^{k} \Delta Y_k$,
    Step 3 Find the first hitting time $\tau_i = \Delta t \cdot \max_{1 \le k \le n} \{k|Y(t_k) \ge h(t_k)\}$.
- Return: Simulated first hitting times $\tau_1, \ldots, \tau_N$

Given the simulated first hitting times $\tau_1, \ldots \tau_N$, one can then obtain the Monte Carlo estimators for the expected failure time $\mathbb{E}[\tau]$ and the failure probability $\Pr(\tau \le t) := p_t$ as

$$\widehat{\mathbb{E}[t]} = \overline{\tau} = \Sigma_{i=1}^{N} \tau_i/N, \tag{8.4}$$

$$\widehat{p}_t = \Sigma_{i=1}^{N} I(\tau_i \le t)/N. \tag{8.5}$$

The above algorithm 8.1 can be easily modified to estimate remaining useful life given the current degradation condition since the inverse Gaussian process is a Markov process. Given the current degradation level $Y(t_0)$ at time t_0, the above algorithm will estimate the RUL after revising the hitting boundary to $\widetilde{h}(t) = h(t + t_0) - Y(t_0)$.

Furthermore, the prediction interval of RUL at the confidence level $0 < \alpha < 1$ can be estimated by the sample $a/2$ quantile and the sample $(1 - a/2)$ quantile as $(\widehat{q}_{a/2}(\tau), \widehat{q}_{1-a/2}(\tau))$.

8.1.3 Estimator Error

This section can further find the MCS standard errors for both two estimators to quantify their uncertainties at the order $O(1/\sqrt{N})$. One shall notice that, from a theoretical perspective, Algorithm 8.1 is simulating $\tau \wedge T$ rather than τ.

For any $\tau_i > T$, only T will be recorded as a right censoring observation. For estimating the failure probability, $\widehat{p}_t$ is still an unbiased estimator if $t \le T$. But $\widehat{\mathbb{E}[t]}$ tends to slightly underestimate the mean time to failure. Nevertheless, one can show that this negative bias induced by right censoring converge to zero at an exponential rate. So, this issue can be resolved by selecting a reasonably large T. Besides the MCS standard error and the bias induced by right censoring, τ_i tends to overestimate the true first hitting time since the boundary crossing happens between t_{k-1} and t_k with t_k being recorded.

This bias induced by the discrete approximation at the order $O(\Delta t)$ can be resolved by choosing a sufficient small Δt. Obviously, the computational complexity of Algorithm 8.1 mainly depends on the MCS sample size N and the number of grid points $K = T/\Delta t$.

The computational complexity can be written as $O(NK) = O(NT/\Delta t)$. Given a fixed number of MCS sample size N and a maximum time limit T, one can choose $\Delta t = T/\sqrt{N}$ to match the order of the MCS standard error with the order of the discrete approximation error. This will result in an algorithm with the computational complexity at $O(N^{3/2})$ and the relative error at $O(1/\sqrt{N})$.

Furthermore, the independence between MCS sample paths and the independence between the increments of each sample path allow us accelerating the computation through vectorisation and parallelization. The real computing time of Algorithm 8.1 is shown to be extremely fast on a standard personal desktop. The details will be presented in the case study of [383].

8.2 Theoretical Stability Analysis

In this section, uncertainties from the gains and deadtime of transfer function of CPSs are discussed and two methods for analyzing the stability of CPSs against time delays are provided.

8.2.1 Impacts of Uncertainties

In the CPS, uncertainties can be associated with the variation in the deadtime L caused by aging process or by transmission delays and to the simplifications used in approximating the dead-time-free dynamics of the control process by a transfer function.

This impact analysis can be made for the general case but this section focuses on the case where the nominal models are the simple first-order-plus-dead-time transfer function, since it is the most widely used in industrial applications [176, 294, 380, 385].

It is demonstrated that the main factor affecting frequency domain uncertainties is the variation in the gains and deadtime [294, 380, 386]. The following uncertainties will be taken into account:

- errors in the estimation of gains K or K_n;
- variation in the deadtime;
- errors in estimating the dominating time constant in the stable case;
- unmodelled dynamics. Because it is customary in most literature, poles and zeros that are faster than the dominate pole can be approximated by a first-order model with an equivalent time constant T_μ, which is smaller than the dominant T [386].

Above uncertainties are commonly found in industrial processes. Therefore, for the stable case the "real" process $P(s)$, which considers both modelling errors and the variation in the deadtime caused by degradation, and the model $P_n(s)$ are represented by

$$P(s) = \frac{K}{(1 + sT)(1 + sT_\mu)} e^{-Ls} \tag{8.6}$$

$$P_n(s) = \frac{K_n}{1 + sT_n} e^{-L_n s} \tag{8.7}$$

where the suffix n represents the nominal value of a parameter or a model. K and K_n are the gains of the control process, and T, T_μ and T_n are the time constants of the control process.

In the stable case, the transfer functions of $P(s)$ and $P_n(s)$ are used to derive the general expression of $\delta P(s)$:

$$\delta P(s) = \frac{\frac{K e^{-Ls}}{(1+sT)(1+sT_\mu)} - \frac{K_n e^{-L_n s}}{1+sT_n}}{\frac{K_n e^{-L_n s}}{1+sT_n}} = \frac{K}{K_n} \frac{(1 + sT_n)e^{-(L-L_n)s}}{(1 + sT)(1 + sT_\mu)} - 1 \tag{8.8}$$

where $\delta P(s)$ is a multiplicative description of the modelling errors or variation in the deadtime.

Using the parameterization of the models with the value of the nominal gain, time constant and variation in the deadtime, (8.8) can be rewritten as

$$\delta P(j\omega_n) = (1 + \delta K) \frac{1 + j\omega_n T_n'}{1 + j\omega_n(T_n' + \delta T)} \frac{1}{1 + j\omega_n T_\mu'} e^{-j\omega_n \delta L} - 1 \tag{8.9}$$

where $\omega_n = \omega L_n$, $L' = L/L_n$, $T' = T/L_n$, $T_\mu' = T_\mu/L_n$ and $T_n' = T_n/L_n$. The incremental (or relative) errors and variation in the deadtime are:

$$\delta L' = \delta L = \frac{L - L_n}{L_n},$$

$$\delta T' = \delta T = \frac{T - T_n}{T_n},$$

$$\delta K = \frac{K - K_n}{K_n}.$$

In order to combine the influence of the modelling errors and the variation in the deadtime, two assumptions are made.

Assumption 8.1 The dead-time-dominant control process is considered, therefore $L \geq T$.

Assumption 8.2 Compared to T, the neglected dynamics represented by T_μ are considered at high frequencies.

From (8.9), it is found that the variation in the deadtime δL has important impact on the robustness of the aging CPS. Thus, the delay margin of the CPS can be defined as the largest variation in the deadtime that is allowed to occur in the process $P(s)$ before the closed-loop CPS becomes unstable, that is

$$1 + C(s)P(s),$$

where $C(s)$ is the controller associated with the process.

8.2.2 Small Gain Theorem based Stability Criteria

The problem of evaluating the stability of linear feedback systems with time-varying but bounded delays was investigated in [387] using the small gain theorem.

The essential idea is to transfer the delay system to be analyzed in such a way that it becomes a feedback interconnection of a linear plant based on the following operator:

$$\Delta_F(x) := \int_{t-d(t)}^{t} x(\theta)d\theta \tag{8.10}$$

where $d(t)$ is the time-varying delay (deadtime) which is assumed to be bounded. Subsequently, an upper bound of the induced gain of the operator Δ_F is computed, which allows the use of the input-output analysis, i.e., the small gain theorem.

The advantage of this approach relies on the fact that the stability criterion is a simple graphical check in a closed-loop Bode plot that makes the design for robustness in different uncertainty delay scenarios convenient.

It is assumed that the closed-loop CPS is stable for zero delay. The following *Theorem* 8.1 presents simple stability criteria of the CPS with arbitrarily time-varying but bounded delays.

Theorem 8.1 For the closed-loop CPS in Figure 3.1 with continuous time and stable plant $G(s)$ as defined in (3.1), and the Proportional–Integral (PI) controller $C(s)$ expressed by (3.5), the CPS is stable for any time-varying delay, and is defined by

$$\Delta(x) = x(t - d(t)), \mathcal{T}_L \leq d(t) \leq \mathcal{T}_U, \tag{8.11}$$

if

$$\left| \frac{\beta G(j\omega)C(j\omega)}{1 + \beta G(j\omega)C(j\omega)} \right| < \frac{1}{\omega \mathcal{T}_U}, \forall \omega \in [0, \infty]. \tag{8.12}$$

The proof of the case with the continuous-time plant and PI controller is provided in [387]. By using the *Theorem* 8.1, the maximum allowable time delay $\mathcal{T}_U$ to maintain the stability of closed-loop CPS can be derived

8.2.3 Robust Stability Criteria

The robust analysis of the closed-loop CPS can be analyzed in a simple way by describing $P(s)$ as

$$P(s) = P_n(s)(1 + \delta P(s)). \tag{8.13}$$

Consider the characteristic closed-loop equation of the aging CPS

$$1 + C(s)P(s) = 1 + C(s)P_n(s)(1 + \delta P(s)) = 0. \tag{8.14}$$

Proposition 8.1: The maximum distance $(\delta P(j\omega))$ between the nominal process $(P_n(s))$ and the real control process $(P(s))$ that maintains the closed-loop robust stability is

$$|\delta P(j\omega)| < dP = \frac{|1 + C(j\omega)P_n(j\omega)|}{|C(j\omega)P_n(j\omega)|}, \forall \omega \geq 0. \tag{8.15}$$

Theorem 8.2 For the closed-loop CPS, the process $P(s)$ with degradation is stable for any variation ΔL in the deadtime if

$$\frac{|C(j\omega)P_n(j\omega)|}{|1 + C(j\omega)P_n(j\omega)|} < \frac{1}{|e^{-j\omega\Delta L} - 1|}, \forall \omega \geq 0 \tag{8.16}$$

where $\Delta L = L - L_n$.

Proof. Note that the $|\delta P(j\omega)|$ in (8.15) is now only determined by the variation in the deadtime. Substituting $K = K_n$, $T = T_n$ and $T_\mu = 0$ into (8.9), $\delta P(j\omega)$ can be computed as

$$|\delta P(j\omega)| = |e^{-j\omega\Delta L} - 1|. \tag{8.17}$$

Therefore, substituting (8.17) into (8.15), the robust stability condition for the closed-loop CPS only with degradation is given as

$$\frac{|C(j\omega)P_n(j\omega)|}{|1 + C(j\omega)P_n(j\omega)|} < \frac{1}{|e^{-j\omega\Delta L} - 1|}, \forall \omega \geq 0. \tag{8.18}$$

By using the *Theorem* 8.2, the maximum allowable variation $\overline{\Delta L}$ in the deadtime to maintain robust stability of the closed-loop CPS can be derived, which is associated with the maximum allowable degradation level in the CPS component.

8.3 Game Model for CPSs

In Chapter 6 and Chapter 7, two different game models are provided to study the cyber security of CPSs. However, they are just two of most common game models. This section discusses other two game models that are also widely used.

Consider the Optimal Power Flow (OPF) model of the smart grid from (7.23) to (7.27). As $s_{i,i'} = 1$ means that power cannot be transmitted between Node i and Node i' due to the successful cyber-attack on Node i, the successful rate can be decided by the vulnerability model in (6.1) given q_i, Q_i and m.

The revised objective function in terms of Expected ENS (EENS) is as follow:

$$\min_{Q,q} EENS = \sum_{l=1}^{2^N} p_l \cdot ENS_l. \tag{8.19}$$

where $l \in \mathbb{N}$ stands for the l-th failure scenario of feeders (N is the number of paired feeders), i.e., $p_l = \prod_{j=1}^{l} v_j \cdot \prod_{j=l+1}^{N}(1 - v_j)$. As a result, ENS_l is the corresponding return value of (7.23) given that the first l feeders fail due to cyber-attacks.

Therefore, the stage I is about the initial attack-defense resource allocation where the defender wants to minimize (8.19) but the attacker wants to maximize (8.19):

$$\min_{q} \max_{Q} \{EENS \| |Q| < \overline{Q}, |q| < \overline{q}, \vartheta_t\} \tag{8.20}$$

where $Q = [Q_1, ..., Q_N]$ and $q = [q_1, ..., q_N]$. (8.20) is a well-defined co-operative game and can be solved using the method proposed in [388].

The stage I is indeed a multi-objective optimization problem and generate K solution.

The stage II is minimizing the maximum EENS obtained from (8.20), which can be formulated as

$$\min_{q} \{EENS | Q', |q| < \overline{q}, \vartheta_t\} \tag{8.21}$$

where $Q' = \arg\max_{1 \le k \le K} EENS_k$ and (8.21) can be solved within $O(1/\epsilon^2)$ iterations of ϵ-subgradient method.

The stage III is dealing with the attacker's response against the optimal defense resource allocation strategy q^* obtained from (8.21) and can be defined as

$$Q^* = \arg\max_{Q} \{EENS | q^*, |Q| < \overline{Q}\} \tag{8.22}$$

where Q^* is the optimal attack resource allocation strategy and can also be solved by the ϵ-subgradient method.

The game between the attacker and defender can also be described by the Nash equilibrium, which can be referred to [389–391].

References

1 Siu, N., 1994. Risk assessment for dynamic systems: an overview. *Reliability Engineering & System Safety*, 43(1), pp. 43–73.

2 Aldemir, T., 2013. A survey of dynamic methodologies for probabilistic safety assessment of nuclear power plants. *Annals of Nuclear Energy*, 52, pp. 113–124.

3 Jiang, K. and Singh, C., 2011. New models and concepts for power system reliability evaluation including protection system failures. *IEEE Transactions on Power Systems*, 4(26), pp. 1845–1855.

4 Levitin, G., 2007. Block diagram method for analyzing multi-state systems with uncovered failures. *Reliability Engineering & System Safety*, 92(6), pp. 727–734.

5 Kim, M.C., 2011. Reliability block diagram with general gates and its application to system reliability analysis. *Annals of Nuclear Energy*, 38(11), pp. 2456–2461.

6 Aldemir, T., Miller, D.W., Stovsky, M., Kirschenbaum, J., Bucci, P., Mangan, L.A., Fentiman, A. and Arndt, S.A., 2007. Methodologies for the probabilistic risk assessment of digital reactor protection and control systems. *Nuclear technology*, 159(2), pp. 167–191.

7 Shin, S.K. and Seong, P.H., 2008. Review of various dynamic modeling methods and development of an intuitive modeling method for dynamic systems. *Nuclear Engineering and Technology*, 40(5), pp. 375–386.

8 Verlinden, S., Deconinck, G., Coupe, B., 2012. Hybrid reliability model for nuclear reactor safety system. *Reliability Engineering and System Safety*, 101, 35–47.

9 Cauffriez, L., Ciccotelli, J., Conrard, B. and Bayart, M., 2004. Design of intelligent distributed control systems: a dependability point of view. *Reliability Engineering & System Safety*, 84(1), pp. 19–32.

10 Clarhaut, J., Conrard, B., Hayat, S. and Cocquempot, V., 2009. Optimal design of dependable control system architectures using temporal sequences of failures. *IEEE Transactions on reliability*, 58(3), pp. 511–522.

11 Cauffriez, L., Benard, V. and Renaux, D., 2006. A new formalism for designing and specifying RAMS parameters for complex distributed control systems: the Safe-SADT formalism. *IEEE Transactions on Reliability*, 55(3), pp. 397–410.

12 Xiao, H., Lee, L.H. and Ng, K.M., 2013. Optimal computing budget allocation for complete ranking. *IEEE Transactions on Automation Science and Engineering*, 11(2), pp. 516–524.

Cyber-Physical Distributed Systems: Modeling, Reliability Analysis and Applications, First Edition.
Huadong Mo, Giovanni Sansavini and Min Xie.
© 2021 John Wiley & Sons Ltd. Published 2021 by John Wiley & Sons Ltd.

13 Ghostine, R., Thiriet, J.M. and Aubry, J.F., 2011. Variable delays and message losses: Influence on the reliability of a control loop. *Reliability Engineering & System Safety*, 96(1), pp. 160–171.

14 Karimi, H.R., Duffie, N.A. and Dashkovskiy, S., 2010. Local capacity H infinity control for production networks of autonomous work systems with time-varying delays. *IEEE Transactions on Automation Science and Engineering*, 7(4), pp. 849–857.

15 Yeh, W.C., Lin, Y.C., Chung, Y.Y. and Chih, M., 2010. A particle swarm optimization approach based on Monte Carlo simulation for solving the complex network reliability problem. *IEEE Transactions on Reliability*, 59(1), pp. 212–221.

16 Kamat, S.J. and Riley, M.W., 1975. Determination of reliability using event-based Monte Carlo simulation. *IEEE transactions on reliability*, 24(1), pp. 73–75.

17 Jiang, L., Yao, W., Wu, Q.H., Wen, J.Y. and Cheng, S.J., 2011. Delay-dependent stability for load frequency control with constant and time-varying delays. *IEEE Transactions on Power systems*, 27(2), pp. 932–941.

18 Tan, W., 2009. Unified tuning of PID load frequency controller for power systems via IMC. *IEEE Transactions on power systems*, 25(1), pp. 341–350.

19 Kim, Y.J., Norford, L.K. and Kirtley, J.L., 2014. Modeling and analysis of a variable speed heat pump for frequency regulation through direct load control. *IEEE Transactions on Power Systems*, 30(1), pp. 397–408.

20 He, Y., Wu, M. and She, J.H., 2006. Delay-dependent stability criteria for linear systems with multiple time delays. *IEE Proceedings-Control Theory and Applications*, 153(4), pp. 447–452.

21 Wu, M., He, Y., She, J.H. and Liu, G.P., 2004. Delay-dependent criteria for robust stability of time-varying delay systems. *Automatica*, 40(8), pp. 1435–1439.

22 Jing, T., Chen, F. and Li, Q., 2015. Finite-time mixed outer synchronization of complex networks with time-varying delay and unknown parameters. *Applied Mathematical Modelling*, 39(23-24), pp. 7734–7743.

23 Zhang, C.K., Jiang, L., Wu, Q.H., He, Y. and Wu, M., 2013. Delay-dependent robust load frequency control for time delay power systems. *IEEE Transactions on Power Systems*, 28(3), pp. 2192–2201.

24 Peng, C. and Zhang, J., 2015. Delay-distribution-dependent load frequency control of power systems with probabilistic interval delays. *IEEE Transactions on Power Systems*, 31(4), pp. 3309–3317.

25 Stahlhut, J.W., Browne, T.J., Heydt, G.T. and Vittal, V., 2008. Latency viewed as a stochastic process and its impact on wide area power system control signals. *IEEE Transactions on Power Systems*, 23(1), pp. 84–91.

26 Upadhyay, R.K. and Kumari, S., 2019. Discrete and data packet delays as determinants of switching stability in wireless sensor networks. *Applied Mathematical Modelling*, 72, pp. 513–536.

27 Joshi, V.V., Xie, L.B., Park, J.J., Shieh, L.S., Chen, Y.H., Grigoriadis, K. and Tsai, J.S.H., 2012. Digital modeling and control of multiple time-delayed distributed power grid. *Applied Mathematical Modelling*, 36(9), pp. 4118–4134.

28 Wu, H., Tsakalis, K.S. and Heydt, G.T., 2004. Evaluation of time delay effects to wide-area power system stabilizer design. *IEEE Transactions on Power Systems*, 19(4), pp. 1935–1941.

29 Wang, J., Liu, C. and Yang, H., 2012. Stability of a class of networked control systems with Markovian characterization. *Applied Mathematical Modelling*, 36(7), pp. 3168–3175.

30 Hu, J.B., Zhao, L.D., Lu, G.P. and Zhang, S.B., 2016. The stability and control of fractional nonlinear system with distributed time delay. *Applied Mathematical Modelling*, 40(4), pp. 3257–3263.

31 Xu, H.T., Zhang, C.K., Jiang, L. and Smith, J., 2017. Stability analysis of linear systems with two additive time-varying delays via delay-product-type Lyapunov functional. *Applied Mathematical Modelling*, 45, pp. 955–964.

32 Zhang, P., Yang, D.Y., Chan, K.W. and Cai, G.W., 2012. Adaptive wide-area damping control scheme with stochastic subspace identification and signal time delay compensation. *IET generation, transmission & distribution*, 6(9), pp. 844–852.

33 Wen, S., Yu, X., Zeng, Z. and Wang, J., 2015. Event-triggering load frequency control for multiarea power systems with communication delays. *IEEE Transactions on Industrial Electronics*, 63(2), pp. 1308–1317.

34 Wang, Y., Xiong, J. and Ren, W., 2018. Decentralised output-feedback LQG control with one-step communication delay. *International Journal of Control*, 91(8), pp. 1920–1930.

35 Rerkpreedapong, D., Hasanovic, A. and Feliachi, A., 2003. Robust load frequency control using genetic algorithms and linear matrix inequalities. *IEEE Transactions on Power Systems*, 18(2), pp. 855–861.

36 Sönmez, Ş., Ayasun, S. and Nwankpa, C.O., 2015. An exact method for computing delay margin for stability of load frequency control systems with constant communication delays. *IEEE Transactions on Power Systems*, 31(1), pp. 370–377.

37 Li, X.G., Niculescu, S.I., Cela, A., Wang, H.H. and Cai, T.Y., 2012. On computing Puiseux series for multiple imaginary characteristic roots of LTI systems with commensurate delays. *IEEE Transactions on Automatic Control*, 58(5), pp. 1338–1343.

38 Walton, K. and Marshall, J.E., 1987, March. Direct method for TDS stability analysis. In *IEE Proceedings D-Control Theory and Applications* (Vol. 134, No. 2, pp. 101–107). IET.

39 Rakkiyappan, R., Lakshmanan, S., Sivasamy, R. and Lim, C.P., 2016. Leakage-delay-dependent stability analysis of Markovian jumping linear systems with time-varying delays and nonlinear perturbations. *Applied Mathematical Modelling*, 40(7-8), pp. 5026–5043.

40 Sun, Y., Li, N., Zhao, X., Wei, Z., Sun, G. and Huang, C., 2016. Robust H∞ load frequency control of delayed multi-area power system with stochastic disturbances. *Neurocomputing*, 193, pp. 58–67.

41 Lai, C.L. and Hsu, P.L., 2009. Design the remote control system with the time-delay estimator and the adaptive smith predictor. *IEEE Transactions on Industrial Informatics*, 6(1), pp. 73–80.

42 Kuzlu, M., Pipattanasomporn, M. and Rahman, S., 2014. Communication network requirements for major smart grid applications in HAN, NAN and WAN. *Computer Networks*, 67, pp. 74–88.

43 Grenier, M. and Navet, N., 2008. Fine-tuning MAC-level protocols for optimized real-time QoS. *IEEE Transactions on Industrial Informatics*, 4(1), pp. 6–15.

44 Yao, J., Liu, X., Zhu, G. and Sha, L., 2012. NetSimplex: Controller fault tolerance architecture in networked control systems. *IEEE Transactions on Industrial Informatics*, 9(1), pp. 346–356.

45 Mahmoud, M.S. and Sabih, M., 2013. Experimental investigations for distributed networked control systems. *IEEE Systems Journal*, 8(3), pp. 717–725.

46 Singh, V.P., Kishor, N. and Samuel, P., 2016. Communication time delay estimation for load frequency control in two-area power system. *Ad Hoc Networks*, 41, pp. 69–85.

47 Yi, J., Wang, Q., Zhao, D. and Wen, J.T., 2007. BP neural network prediction-based variable-period sampling approach for networked control systems. *Applied Mathematics and Computation*, 185(2), pp. 976–988.

48 Gil-González, W., Garces, A. and Escobar, A., 2019. Passivity-based control and stability analysis for hydro-turbine governing systems. *Applied Mathematical Modelling*, 68, pp. 471–486.

49 Truong, D.Q., Ahn, K.K. and Trung, N.T., 2012. Design of an advanced time delay measurement and a smart adaptive unequal interval grey predictor for real-time nonlinear control systems. *IEEE Transactions on Industrial Electronics*, 60(10), pp. 4574–4589.

50 Han, C. and Zhang, H., 2009. Linear optimal filtering for discrete-time systems with random jump delays. *Signal Processing*, 89(6), pp. 1121–1128.

51 Evans, J.S. and Krishnamurthy, V., 1999. Hidden Markov model state estimation with randomly delayed observations. *IEEE Transactions on Signal Processing*, 47(8), pp. 2157–2166.

52 Ge, Y., Chen, Q., Jiang, M. and Huang, Y., 2013. Modeling of random delays in networked control systems. *Journal of Control Science and Engineering, 2013*.

53 Narbutt, M. and Murphy, L., 2003. VoIP playout buffer adjustment using adaptive estimation of network delays. In *Teletraffic Science and Engineering* (Vol. 5, pp. 1171–1180). Elsevier.

54 J. Nilsson, *Real-Time Control Systems with Delays*, 1998, Lund Institute of Technology.

55 Lai, C.L. and Hsu, P.L., 2014. The butterfly-shaped feedback loop in networked control systems for the unknown delay compensation. *IEEE Transactions on Industrial Informatics*, 10(3), pp. 1746–1754.

56 Huang, D. and Nguang, S.K., 2008. State feedback control of uncertain networked control systems with random time delays. *IEEE Transactions on automatic control*, 53(3), pp. 829–834.

57 Cordova-Garcia, J., Wang, X., Xie, D., Zhao, Y. and Zuo, L., 2018. Control of Communications-Dependent Cascading Failures in Power Grids. *IEEE Transactions on Smart Grid*, 10(5), pp. 5021–5031.

58 Cong, S., Ge, Y., Chen, Q., Jiang, M. and Shang, W., 2010. DTHMM based delay modeling and prediction for networked control systems. *Journal of Systems Engineering and Electronics*, 21(6), pp. 1014–1024.

59 Ge, Y., Chen, Q., Jiang, M. and Huang, Y., 2014. SCHMM-based modeling and prediction of random delays in networked control systems. *Journal of The Franklin Institute*, 351(5), pp. 2430–2453.

60 Ge, Y., Zhang, X., Chen, Q. and Jiang, M., 2016. Initialization of the HMM-based delay model in networked control systems. *Information Sciences*, 364, pp. 1–15.

61 Huang, D. and Nguang, S.K., 2008. Robust disturbance attenuation for uncertain networked control systems with random time delays. *IET Control Theory & Applications*, 2(11), pp. 1008–1023.

62 Huang, Q., Shao, L. and Li, N., 2015. Dynamic detection of transmission line outages using hidden Markov models. *IEEE Transactions on power systems*, 31(3), pp. 2026–2033.

63 Tabone, M.D. and Callaway, D.S., 2014. Modeling variability and uncertainty of photovoltaic generation: A hidden state spatial statistical approach. *IEEE Transactions on Power Systems*, 30(6), pp. 2965–2973.

64 Albert, A. and Rajagopal, R., 2013. Smart meter driven segmentation: What your consumption says about you. *IEEE Transactions on power systems*, 28(4), pp. 4019–4030.

65 Albert, A. and Rajagopal, R., 2014. Thermal profiling of residential energy use. *IEEE Transactions on power systems*, 30(2), pp. 602–611.

66 Guo, Z., Wang, Z.J. and Kashani, A., 2014. Home appliance load modeling from aggregated smart meter data. *IEEE Transactions on power systems*, 30(1), pp. 254–262.

67 Welch, L.R., 2003. Hidden Markov models and the Baum-Welch algorithm. *IEEE Information Theory Society Newsletter*, 53(4), pp. 10–13.

68 Viterbi, A.J. and Omura, J.K., 2013. *Principles of digital communication and coding.* Courier Corporation.

69 Chaudhuri, B., Majumder, R. and Pal, B.C., 2004. Wide-area measurement-based stabilizing control of power system considering signal transmission delay. *IEEE Transactions on Power Systems*, 19(4), pp. 1971–1979.

70 Driesen, J. and Katiraei, F., 2008. Design for distributed energy resources. *IEEE power and energy magazine*, 6(3), pp. 30–40.

71 Basak, P., Chowdhury, S., nee Dey, S.H. and Chowdhury, S.P., 2012. A literature review on integration of distributed energy resources in the perspective of control, protection and stability of microgrid. *Renewable and Sustainable Energy Reviews*, 16(8), pp. 5545–5556.

72 Kumar, L.S., Kumar, G.N. and Madichetty, S., 2017. Pattern search algorithm based automatic online parameter estimation for AGC with effects of wind power. *International Journal of Electrical Power & Energy Systems*, 84, pp. 135–142.

73 Shotorbani, A.M., Zadeh, S.G., Mohammadi-Ivatloo, B. and Hosseini, S.H., 2017. A distributed non-Lipschitz control framework for self-organizing microgrids with uncooperative and renewable generations. *International Journal of Electrical Power & Energy Systems*, 90, pp. 267–279.

74 Das, D.C., Roy, A.K. and Sinha, N., 2012. GA based frequency controller for solar thermal–diesel–wind hybrid energy generation/energy storage system. *International Journal of Electrical Power & Energy Systems*, 43(1), pp. 262–279.

75 Dash, P., Saikia, L.C. and Sinha, N., 2016. Flower pollination algorithm optimized PI-PD cascade controller in automatic generation control of a multi-area power system. *International Journal of Electrical Power & Energy Systems*, 82, pp. 19–28.

76 Shiva, C.K. and Mukherjee, V., 2016. Design and analysis of multi-source multi-area deregulated power system for automatic generation control using quasi-oppositional harmony search algorithm. *International Journal of Electrical Power & Energy Systems*, 80, pp. 382–395.

77 Lee, D.J. and Wang, L., 2008. Small-signal stability analysis of an autonomous hybrid renewable energy power generation/energy storage system part I: Time-domain simulations. *IEEE Transactions on energy conversion*, 23(1), pp. 311–320.

78 Khadanga, R.K. and Satapathy, J.K., 2015. Time delay approach for PSS and SSSC based coordinated controller design using hybrid PSO–GSA algorithm. *International Journal of Electrical Power & Energy Systems*, 71, pp. 262–273.

79 Pan, I. and Das, S., 2015. Fractional order AGC for distributed energy resources using robust optimization. *IEEE transactions on smart grid*, 7(5), pp. 2175–2186.

80 Liu, S., Liu, P.X. and El Saddik, A., 2014. Modeling and stability analysis of automatic generation control over cognitive radio networks in smart grids. *IEEE Transactions on Systems, Man, and Cybernetics: Systems*, 45(2), pp. 223–234.

81 Borghetti, A., Bottura, R., Barbiroli, M. and Nucci, C.A., 2016. Synchrophasors-based distributed secondary voltage/VAR control via cellular network. *IEEE Transactions on Smart Grid*, 8(1), pp. 262–274.

82 Timbus, A., Larsson, M. and Yuen, C., 2009. Active management of distributed energy resources using standardized communications and modern information technologies. *IEEE Transactions on Industrial Electronics*, 56(10), pp. 4029–4037.

83 Rakhshani, E., Remon, D. and Rodriguez, P., 2016. Effects of PLL and frequency measurements on LFC problem in multi-area HVDC interconnected systems. *International Journal of Electrical Power & Energy Systems*, 81, pp. 140–152.

84 Han, G., Xu, B., Fan, K. and Lv, G., 2014. An open communication architecture for distribution automation based on IEC 61850. *International Journal of Electrical Power & Energy Systems*, 54, pp. 315–324.

85 Ahmadi, A. and Aldeen, M., 2017. Robust overlapping load frequency output feedback control of multi-area interconnected power systems. *International Journal of Electrical Power & Energy Systems*, 89, pp. 156–172.

86 Baghaee, H.R., Mirsalim, M., Gharehpetian, G.B. and Talebi, H.A., 2017. A generalized descriptor-system robust H∞ control of autonomous microgrids to improve small and large signal stability considering communication delays and load nonlinearities. *International Journal of Electrical Power & Energy Systems*, 92, pp. 63–82.

87 Abdelaziz, A.Y. and Ali, E.S., 2015. Cuckoo search algorithm based load frequency controller design for nonlinear interconnected power system. *International Journal of Electrical Power & Energy Systems*, 73, pp. 632–643.

88 Peng, C. and Han, Q.L., 2015. On designing a novel self-triggered sampling scheme for networked control systems with data losses and communication delays. *IEEE Transactions on Industrial Electronics*, 63(2), pp. 1239–1248.

89 Mo, H.D., Li, Y.F. and Zio, E., 2016. A system-of-systems framework for the reliability analysis of distributed generation systems accounting for the impact of degraded communication networks. *Applied Energy*, 183, pp. 805–822.

90 Gungor, V.C., Sahin, D., Kocak, T., Ergut, S., Buccella, C., Cecati, C. and Hancke, G.P., 2012. A survey on smart grid potential applications and communication requirements. *IEEE Transactions on industrial informatics*, 9(1), pp. 28–42.

91 Lu, X., Yu, X., Lai, J., Guerrero, J.M. and Zhou, H., 2016. Distributed secondary voltage and frequency control for islanded microgrids with uncertain communication links. *IEEE Transactions on Industrial Informatics*, 13(2), pp. 448–460.

92 Wang, W. and Lu, Z., 2013. Cyber security in the smart grid: Survey and challenges. *Computer networks*, 57(5), pp. 1344–1371.

93 Tian, G., Camtepe, S. and Tian, Y.C., 2016. A deadline-constrained 802.11 MAC protocol with QoS differentiation for soft real-time control. *IEEE Transactions on Industrial Informatics*, 12(2), pp. 544–554.

94 Padilla E, Agbossou K, Cardenas A. Towards smart integration of distributed energy resources using distributed network protocol over Ethernet. *IEEE Trans Smart Grid* 2014;5(4):1686–93.

95 Zurawski R. ed., 2014. *Industrial communication technology handbook*. CRC Press.

96 Ghoshal, S.P., 2004. Optimizations of PID gains by particle swarm optimizations in fuzzy based automatic generation control. *Electric Power Systems Research*, 72(3), pp. 203–212.

97 Baghaee, H.R., Mirsalim, M. and Gharehpetian, G.B., 2016. Performance improvement of multi-DER microgrid for small-and large-signal disturbances and nonlinear loads: Novel complementary control loop and fuzzy controller in a hierarchical droop-based control scheme. *IEEE Systems Journal*, 12(1), pp. 444–451.

98 Baghaee, H.R., Mirsalim, M. and Gharehpetian, G.B., 2016. Real-time verification of new controller to improve small/large-signal stability and fault ride-through capability of multi-DER microgrids. *IET Generation, Transmission & Distribution*, 10(12), pp. 3068–3084.

99 Baghaee, H.R., Mirsalim, M., Gharehpetian, G.B. and Talebi, H.A., 2017. A decentralized power management and sliding mode control strategy for hybrid AC/DC microgrids including renewable energy resources. *IEEE transactions on industrial informatics.*

100 Guerrero, J.M., Loh, P.C., Lee, T.L. and Chandorkar, M., 2012. Advanced control architectures for intelligent microgrids—Part II: Power quality, energy storage, and AC/DC microgrids. *IEEE Transactions on industrial electronics*, 60(4), pp. 1263–1270.

101 Etemadi, A.H., Davison, E.J. and Iravani, R., 2014. A generalized decentralized robust control of islanded microgrids. *IEEE transactions on Power Systems*, 29(6), pp. 3102–3113.

102 Babazadeh, M. and Karimi, H., 2013. A robust two-degree-of-freedom control strategy for an islanded microgrid. *IEEE transactions on power delivery*, 28(3), pp. 1339–1347.

103 Baghaee, H.R., Mirsalim, M. and Gharehpetian, G.B., 2016. Power calculation using RBF neural networks to improve power sharing of hierarchical control scheme in multi-DER microgrids. *IEEE Journal of Emerging and Selected Topics in Power Electronics*, 4(4), pp. 1217–1225.

104 Baghaee, H.R., Mirsalim, M., Gharehpetan, G.B. and Talebi, H.A., 2017. Nonlinear load sharing and voltage compensation of microgrids based on harmonic power-flow calculations using radial basis function neural networks. *IEEE systems journal*, 12(3), pp. 2749–2759.

105 Baghaee, H.R., Mirsalim, M., Gharehpetian, G.B. and Talebi, H.A., 2017. Three-phase AC/DC power-flow for balanced/unbalanced microgrids including wind/solar, droop-controlled and electronically-coupled distributed energy resources using radial basis function neural networks. *IET Power Electronics*, 10(3), pp. 313–328.

106 Baghaee, H.R., Mirsalim, M., Gharehpetian, G.B. and Talebi, H.A., 2017. Eigenvalue, robustness and time delay analysis of hierarchical control scheme in multi-DER microgrid to enhance small/large-signal stability using complementary loop and fuzzy logic controller. *Journal of Circuits, Systems and Computers*, 26(06), p. 1750099.

107 Kahrobaeian, A. and Mohamed, Y.A.R.I., 2014. Networked-based hybrid distributed power sharing and control for islanded microgrid systems. *IEEE Transactions on Power Electronics*, 30(2), pp. 603–617.

108 Lai, J., Zhou, H., Lu, X., Yu, X. and Hu, W., 2016. Droop-based distributed cooperative control for microgrids with time-varying delays. *IEEE Transactions on Smart Grid*, 7(4), pp. 1775–1789.

109 Dörfler, F., Jovanović, M.R., Chertkov, M. and Bullo, F., 2014. Sparsity-promoting optimal wide-area control of power networks. *IEEE Transactions on Power Systems*, 29(5), pp. 2281–2291.

110 Wang, J., Conejo, A.J., Wang, C. and Yan, J., 2012. Smart grids, renewable energy integration, and climate change mitigation-Future electric energy systems. *Applied Energy*, 96, pp. 1–3.

111 Desideri, U. and Yan, J., 2012. Clean energy technologies and systems for a sustainable world.

112 Manfren, M., Caputo, P. and Costa, G., 2011. Paradigm shift in urban energy systems through distributed generation: Methods and models. *Applied Energy*, 88(4), pp. 1032–1048.

113 Demirbas, A., 2009. Political, economic and environmental impacts of biofuels: A review. *Applied energy*, 86, pp. 108–117.

114 Kalantar, M., 2010. Dynamic behavior of a stand-alone hybrid power generation system of wind turbine, microturbine, solar array and battery storage. *Applied energy*, 87(10), pp. 3051–3064.

115 Morales, J.M., Minguez, R. and Conejo, A.J., 2010. A methodology to generate statistically dependent wind speed scenarios. *Applied Energy*, 87(3), pp. 843–855.

116 Mahmud, K. and Town, G.E., 2016. A review of computer tools for modeling electric vehicle energy requirements and their impact on power distribution networks. *Applied Energy*, 172, pp. 337–359.

117 Guerra, O.J., Tejada, D.A. and Reklaitis, G.V., 2016. An optimization framework for the integrated planning of generation and transmission expansion in interconnected power systems. *Applied energy*, 170, pp. 1–21.

118 Parra, D., Norman, S.A., Walker, G.S. and Gillott, M., 2016. Optimum community energy storage system for demand load shifting. *Applied Energy*, 174, pp. 130–143.

119 Tahir, M. and Mazumder, S.K., 2015. Self-triggered communication enabled control of distributed generation in microgrids. *IEEE Transactions on Industrial Informatics*, 11(2), pp. 441–449.

120 Keane, A., Ochoa, L.F., Borges, C.L., Ault, G.W., Alarcon-Rodriguez, A.D., Currie, R.A., Pilo, F., Dent, C. and Harrison, G.P., 2012. State-of-the-art techniques and challenges ahead for distributed generation planning and optimization. *IEEE Transactions on Power Systems*, 28(2), pp. 1493–1502.

121 Mohammed, Y.S., Mustafa, M.W. and Bashir, N., 2014. Hybrid renewable energy systems for off-grid electric power: Review of substantial issues. *Renewable and Sustainable Energy Reviews*, 35, pp. 527–539.

122 Ellabban, O., Abu-Rub, H. and Blaabjerg, F., 2014. Renewable energy resources: Current status, future prospects and their enabling technology. *Renewable and Sustainable Energy Reviews*, 39, pp. 748–764.

123 Zeng, B., Zhang, J., Yang, X., Wang, J., Dong, J. and Zhang, Y., 2013. Integrated planning for transition to low-carbon distribution system with renewable energy generation and demand response. *IEEE Transactions on Power Systems*, 29(3), pp. 1153–1165.

124 Gill, S., Kockar, I. and Ault, G.W., 2013. Dynamic optimal power flow for active distribution networks. *IEEE Transactions on Power Systems*, 29(1), pp. 121–131.

125 Purchala, K., Meeus, L., Van Dommelen, D. and Belmans, R., 2005, June. Usefulness of DC power flow for active power flow analysis. In *IEEE Power Engineering Society General Meeting, 2005* (pp. 454–459). IEEE.

126 Baki, A.K.M., 2014. Continuous monitoring of smart grid devices through multi protocol label switching. *IEEE Transactions on Smart Grid*, 5(3), pp. 1210–1215.

127 Mudumbai, R., Dasgupta, S. and Cho, B.B., 2012. Distributed control for optimal economic dispatch of a network of heterogeneous power generators. *IEEE Transactions on Power Systems*, 27(4), pp. 1750–1760.

128 Ancillotti, E., Bruno, R. and Conti, M., 2013. The role of communication systems in smart grids: Architectures, technical solutions and research challenges. *Computer Communications*, 36(17-18), pp. 1665–1697.

129 Niyato, D. and Wang, P., 2012. Cooperative transmission for meter data collection in smart grid. *IEEE Communications Magazine*, 50(4), pp. 90–97.

130 Guo, S., Li, H., Zhao, J., Li, X. and Yan, J., 2013. Numerical simulation study on optimizing charging process of the direct contact mobilized thermal energy storage. *Applied energy*, 112, pp. 1416–1423.

131 Xu, Q., Mak, T., Ko, J. and Sengupta, R., 2007. Medium access control protocol design for vehicle–vehicle safety messages. *IEEE Transactions on Vehicular Technology*, 56(2), pp. 499–518.

132 Meng, L., Zhao, X., Tang, F., Savaghebi, M., Dragicevic, T., Vasquez, J.C. and Guerrero, J.M., 2015. Distributed voltage unbalance compensation in islanded microgrids by using a dynamic consensus algorithm. *IEEE Transactions on Power Electronics*, 31(1), pp. 827–838.

133 Guerrieri, L., Masera, G., Stievano, I.S., Bisaglia, P., Valverde, W.R.G. and Concolato, M., 2016. Automotive power-line communication channels: mathematical characterization and hardware emulator. *IEEE Transactions on Industrial Electronics*, 63(5), pp. 3081–3090.

134 Lin, H., Veda, S.S., Shukla, S.S., Mili, L. and Thorp, J., 2012. GECO: Global event-driven co-simulation framework for interconnected power system and communication network. *IEEE Transactions on Smart Grid*, 3(3), pp. 1444–1456.

135 Gungor, V.C., Sahin, D., Kocak, T., Ergut, S., Buccella, C., Cecati, C. and Hancke, G.P., 2011. Smart grid technologies: Communication technologies and standards. *IEEE transactions on Industrial informatics*, 7(4), pp. 529–539.

136 Tomsovic, K., Bakken, D.E., Venkatasubramanian, V. and Bose, A., 2005. Designing the next generation of real-time control, communication, and computations for large power systems. *Proceedings of the IEEE*, 93(5), pp. 965–979.

137 Bose, A., 2010. Smart transmission grid applications and their supporting infrastructure. *IEEE Transactions on Smart Grid*, 1(1), pp. 11–19.

138 Yu, F.R., Zhang, P., Xiao, W. and Choudhury, P., 2011. Communication systems for grid integration of renewable energy resources. *IEEE network*, 25(5), pp. 22–29.

139 Yu, F.R., Zhang, P., Xiao, W. and Choudhury, P., 2011. Communication systems for grid integration of renewable energy resources. *IEEE network*, 25(5), pp. 22–29.

140 Wu, J. and Chen, T., 2007. Design of networked control systems with packet dropouts. *IEEE Transactions on Automatic control*, 52(7), pp. 1314–1319.

141 Niyato, D., Wang, P., Han, Z. and Hossain, E., 2011, March. Impact of packet loss on power demand estimation and power supply cost in smart grid. In *2011 IEEE Wireless Communications and Networking Conference* (pp. 2024–2029). IEEE.

142 Ruiz-Romero, S., Colmenar-Santos, A., Mur-Pérez, F. and López-Rey, Á., 2014. Integration of distributed generation in the power distribution network: The need for smart grid control systems, communication and equipment for a smart city—Use cases. *Renewable and sustainable energy reviews*, 38, pp. 223–234.

143 Reddy, K.S., Kumar, M., Mallick, T.K., Sharon, H. and Lokeswaran, S., 2014. A review of Integration, Control, Communication and Metering (ICCM) of renewable energy based smart grid. *Renewable and Sustainable Energy Reviews*, 38, pp. 180–192.

144 Dotta, D., e Silva, A.S. and Decker, I.C., 2008. Wide-area measurements-based two-level control design considering signal transmission delay. *IEEE Transactions on Power Systems*, 24(1), pp. 208–216.

145 Liu, S., Liu, X.P. and El Saddik, A., 2014. Modeling and distributed gain scheduling strategy for load frequency control in smart grids with communication topology changes. *ISA transactions*, 53(2), pp. 454–461.

146 Knapp, E.D. and Langill, J.T., 2014. *Industrial Network Security: Securing critical infrastructure networks for smart grid, SCADA, and other Industrial Control Systems.* Syngress.

147 Jacobs, K., 2010. *Stochastic processes for physicists: understanding noisy systems.* Cambridge University Press.

148 Zhang, H., Shi, Y. and Wang, J., 2013. Observer-based tracking controller design for networked predictive control systems with uncertain Markov delays. *International Journal of Control*, 86(10), pp. 1824–1836.

149 Wang, Z., Liu, Y. and Liu, X., 2010. Exponential stabilization of a class of stochastic system with Markovian jump parameters and mode-dependent mixed time-delays. *IEEE Transactions on Automatic Control*, 55(7), pp. 1656–1662.

150 Pandey, S.K., Mohanty, S.R. and Kishor, N., 2013. A literature survey on load–frequency control for conventional and distribution generation power systems. *Renewable and Sustainable Energy Reviews*, 25, pp. 318–334.

151 Majumder, R., Bag, G. and Kim, K.H., 2012. Power sharing and control in distributed generation with wireless sensor networks. *IEEE Transactions on Smart Grid*, 3(2), pp. 618–634.

152 Jardine, A.K., Lin, D. and Banjevic, D., 2006. A review on machinery diagnostics and prognostics implementing condition-based maintenance. *Mechanical systems and signal processing*, 20(7), pp. 1483–1510.

153 Li, Y., Zuo, M.J., Lin, J. and Liu, J., 2017. Fault detection method for railway wheel flat using an adaptive multiscale morphological filter. *Mechanical Systems and Signal Processing*, 84, pp. 642–658.

154 J. Liu, Y.F. Li, and E. Zio, "A SVM framework for fault detection of the braking system in a high speed train," *Mechan. Syst. Signal Process.*, vol. 87, pp. 401–409, Mar. 2017.

155 Pham, H.T., Yang, B.S. and Nguyen, T.T., 2012. Machine performance degradation assessment and remaining useful life prediction using proportional hazard model and support vector machine. *Mechanical Systems and Signal Processing*, 32, pp. 320–330.

156 Lin, Y.H., Li, Y.F. and Zio, E., 2015. A reliability assessment framework for systems with degradation dependency by combining binary decision diagrams and Monte Carlo simulation. *IEEE Transactions on Systems, Man, and Cybernetics: Systems*, 46(11), pp. 1556–1564.

157 Chen, B., Niu, Y. and Zou, Y., 2013. Adaptive sliding mode control for stochastic Markovian jumping systems with actuator degradation. *Automatica*, 49(6), pp. 1748–1754.

158 Jiang, H., Chen, J. and Dong, G., 2016. Hidden Markov model and nuisance attribute projection based bearing performance degradation assessment. *Mechanical systems and signal processing*, 72, pp. 184–205.

159 Li, J., Wang, Z., Zhang, Y., Fu, H., Liu, C. and Krishnaswamy, S., 2017. Degradation data analysis based on a generalized Wiener process subject to measurement error. *Mechanical Systems and Signal Processing*, 94, pp. 57–72.

160 Debbarma, S. and Dutta, A., 2016. Utilizing electric vehicles for LFC in restructured power systems using fractional order controller. *IEEE transactions on smart grid*, 8(6), pp. 2554–2564.

161 Li, Y.G. and Nilkitsaranont, P., 2009. Gas turbine performance prognostic for condition-based maintenance. *Applied energy*, 86(10), pp. 2152–2161.

162 Becejac, T., Dehghanian, P. and Kezunovic, M., 2016, October. Probabilistic assessment of PMU integrity for planning of periodic maintenance and testing. In *2016 International Conference on Probabilistic Methods Applied to Power Systems (PMAPS)* (pp. 1–6). IEEE.

163 Peng, C.Y. and Tseng, S.T., 2009. Mis-specification analysis of linear degradation models. *IEEE Transactions on Reliability*, 58(3), pp. 444–455.

164 STANDARD, B. and ISO, B., 2006. *Reciprocating internal combustion engine driven alternating current generating sets— (Doctoral dissertation, Institute of Technology Tallaght).*

165 Sun, J., Zuo, H., Wang, W. and Pecht, M.G., 2012. Application of a state space modeling technique to system prognostics based on a health index for condition-based maintenance. *Mechanical Systems and Signal Processing*, 28, pp. 585–596.

166 Beganovic, N. and Söffker, D., 2017. Remaining lifetime modeling using State-of-Health estimation. *Mechanical Systems and Signal Processing*, 92, pp. 107–123.

167 Caballé, N.C., Castro, I.T., Pérez, C.J. and Lanza-Gutiérrez, J.M., 2015. A condition-based maintenance of a dependent degradation-threshold-shock model in a system with multiple degradation processes. *Reliability Engineering & System Safety*, 134, pp. 98–109.

168 Si, X.S. and Zhou, D., 2013. A generalized result for degradation model-based reliability estimation. *IEEE Transactions on Automation Science and Engineering*, 11(2), pp. 632–637.

169 Liu, K. and Huang, S., 2014. Integration of data fusion methodology and degradation modeling process to improve prognostics. *IEEE Transactions on Automation Science and Engineering*, 13(1), pp. 344–354.

170 Si, X.S., Wang, W., Hu, C.H. and Zhou, D.H., 2011. Remaining useful life estimation–a review on the statistical data driven approaches. *European journal of operational research*, 213(1), pp. 1–14.

171 Rougé, C., Mathias, J.D. and Deffuant, G., 2014. Relevance of control theory to design and maintenance problems in time-variant reliability: The case of stochastic viability. *Reliability Engineering & System Safety*, 132, pp. 250–260.

172 Si, X.S., Wang, W., Hu, C.H., Chen, M.Y. and Zhou, D.H., 2013. A Wiener-process-based degradation model with a recursive filter algorithm for remaining useful life estimation. *Mechanical Systems and Signal Processing*, 35(1-2), pp. 219–237.

173 Liu, B., Xu, Z., Xie, M. and Kuo, W., 2014. A value-based preventive maintenance policy for multi-component system with continuously degrading components. *Reliability Engineering & System Safety*, 132, pp. 83–89.

174 Huang, Z., Xu, Z., Ke, X., Wang, W. and Sun, Y., 2017. Remaining useful life prediction for an adaptive skew-Wiener process model. *Mechanical Systems and Signal Processing*, 87, pp. 294–306.

175 Mo, H. and Xie, M., 2015. A dynamic approach to performance analysis and reliability improvement of control systems with degraded components. *IEEE Transactions on Systems, Man, and Cybernetics: Systems*, 46(10), pp. 1404–1414.

176 Langeron, Y., Grall, A. and Barros, A., 2015. A modeling framework for deteriorating control system and predictive maintenance of actuators. *Reliability Engineering & System Safety*, 140, pp. 22–36.

177 Crowder, M. and Lawless, J., 2007. On a scheme for predictive maintenance. *European Journal of Operational Research*, 176(3), pp. 1713–1722.

178 Gebraeel, N.Z., Lawley, M.A., Li, R. and Ryan, J.K., 2005. Residual-life distributions from component degradation signals: A Bayesian approach. *IIE Transactions*, 37(6), pp. 543–557.

179 Si, X.S., Wang, W., Chen, M.Y., Hu, C.H. and Zhou, D.H., 2013. A degradation path-dependent approach for remaining useful life estimation with an exact and closed-form solution. *European Journal of Operational Research*, 226(1), pp. 53–66.

180 Si, X.S., Wang, W., Hu, C.H. and Zhou, D.H., 2014. Estimating remaining useful life with three-source variability in degradation modeling. *IEEE Transactions on Reliability*, 63(1), pp. 167–190.

181 Tang, J. and Su, T.S., 2008. Estimating failure time distribution and its parameters based on intermediate data from a Wiener degradation model. *Naval Research Logistics (NRL)*, 55(3), pp. 265–276.

182 Tang, J. and Su, T.S., 2008. Estimating failure time distribution and its parameters based on intermediate data from a Wiener degradation model. *Naval Research Logistics (NRL)*, 55(3), pp. 265–276.

183 Hu, Y., Baraldi, P., Di Maio, F. and Zio, E., 2017. A Systematic Semi-Supervised Self-adaptable Fault Diagnostics approach in an evolving environment. *Mechanical Systems and Signal Processing*, 88, pp. 413–427.

184 Wang, X., 2010. Wiener processes with random effects for degradation data. *Journal of Multivariate Analysis*, 101(2), pp. 340–351.

185 Ye, Z.S., Chen, N. and Shen, Y., 2015. A new class of Wiener process models for degradation analysis. *Reliability Engineering & System Safety*, 139, pp. 58–67.

186 Peng, C.Y., 2015. Inverse Gaussian processes with random effects and explanatory variables for degradation data. *Technometrics*, 57(1), pp. 100–111.

187 Bier, V.M., Nagaraj, A. and Abhichandani, V., 2005. Protection of simple series and parallel systems with components of different values. *Reliability Engineering & System Safety*, 87(3), pp. 315–323.

188 Zhang, C. and Ramirez-Marquez, J.E., 2013. Protecting critical infrastructures against intentional attacks: A two-stage game with incomplete information. *IIE Transactions*, 45(3), pp. 244–258.

189 Nikoofal, M.E. and Zhuang, J., 2015. On the value of exposure and secrecy of defense system: First-mover advantage vs. robustness. *European Journal of Operational Research*, 246(1), pp. 320–330.

190 Xu, Z., Ji, Y. and Zhou, D., 2009. A new real-time reliability prediction method for dynamic systems based on on-line fault prediction. *IEEE transactions on reliability*, 58(3), pp. 523–538.

191 Levitin, G. and Hausken, K., 2009. Redundancy vs. protection in defending parallel systems against unintentional and intentional impacts. *IEEE transactions on reliability*, 58(4), pp. 679–690.

192 Levitin, G. and Hausken, K., 2011. Defense resource distribution between protection and redundancy for constant resource stockpiling pace. *Risk Analysis: An International Journal*, 31(10), pp. 1632–1645.

193 Ramirez-Marquez, J.E. and Rocco, C.M., 2012. Vulnerability based robust protection strategy selection in service networks. *Computers & Industrial Engineering*, 63(1), pp. 235–242.

194 Haphuriwat, N. and Bier, V.M., 2011. Trade-offs between target hardening and overarching protection. *European Journal of Operational Research*, 213(1), pp. 320–328.

195 Perea, F. and Puerto, J., 2013. Revisiting a game theoretic framework for the robust railway network design against intentional attacks. *European Journal of Operational Research*, 226(2), pp. 286–292.

196 Zhuang, J. and Bier, V.M., 2007. Balancing terrorism and natural disasters - Defensive strategy with endogenous attacker effort. *Operations Research*, 55(5), pp. 976–991.

197 Bilis, E.I., Kröger, W. and Nan, C., 2013. Performance of electric power systems under physical malicious attacks. *IEEE Systems Journal*, 7(4), pp. 854–865.

198 Qiao, J., Jeong, D., Lawley, M., Richard, J.P.P., Abraham, D.M. and Yih, Y., 2007. Allocating security resources to a water supply network. *IIE Transactions*, 39(1), pp. 95–109.

199 Levitin, G. and Ben-Haim, H., 2008. Importance of protections against intentional attacks. *Reliability Engineering & System Safety*, 93(4), pp. 639–646.

200 Hausken, K. and Levitin, G., 2009. Minmax defense strategy for complex multi-state systems. *Reliability Engineering & System Safety*, 94(2), pp. 577–587.

201 Hausken, K. and Zhuang, J., 2011. Governments' and terrorists' defense and attack in a T-period game. *Decision Analysis*, 8(1), pp. 46–70.

202 Guikema, S.D. and Aven, T., 2010. Assessing risk from intelligent attacks: A perspective on approaches. *Reliability Engineering & System Safety*, 95(5), pp. 478–483.

203 Peng, R., Levitin, G., Xie, M. and Ng, S.H., 2010. Defending simple series and parallel systems with imperfect false targets. *Reliability Engineering & System Safety*, 95(6), pp. 679–688.

204 Wang, L., Ren, S., Korel, B., Kwiat, K.A. and Salerno, E., 2013. Improving system reliability against rational attacks under given resources. *IEEE Transactions on Systems, Man, and Cybernetics: Systems*, 44(4), pp. 446–456.

205 Levitin, G., Hausken, K. and Dai, Y., 2014. Optimal defense with variable number of overarching and individual protections. *Reliability Engineering & System Safety*, 123, pp. 81–90.

206 Mo, H., Xie, M. and Levitin, G., 2015. Optimal resource distribution between protection and redundancy considering the time and uncertainties of attacks. *European Journal of Operational Research*, 243(1), pp. 200–210.

207 Bricha, N. and Nourelfath, M., 2013. Critical supply network protection against intentional attacks: A game-theoretical model. *Reliability Engineering & System Safety*, 119, pp. 1–10.

208 Zhang, C., Ramirez-Marquez, J.E. and Wang, J., 2015. Critical infrastructure protection using secrecy–A discrete simultaneous game. *European Journal of Operational Research*, 242(1), pp. 212–221.

209 Peng, R., Zhai, Q.Q. and Levitin, G., 2016. Defending a single object against an attacker trying to detect a subset of false targets. *Reliability Engineering & System Safety*, 149, pp. 137–147.

210 Hausken, K. and He, F., 2016. On the effectiveness of security countermeasures for critical infrastructures. *Risk Analysis*, 36(4), pp. 711–726.

211 Ramirez-Marquez, J.E., Rocco, C.M. and Levitin, G., 2011. Optimal network protection against diverse interdictor strategies. *Reliability Engineering & System Safety*, 96(3), pp. 374–382.

212 Ouyang, M., Zhao, L., Hong, L. and Pan, Z., 2014. Comparisons of complex network based models and real train flow model to analyze Chinese railway vulnerability. *Reliability Engineering & System Safety*, 123, pp. 38–46.

213 Xie, G., Hei, X., Mochizuki, H., Takahashi, S. and Nakamura, H., 2014. Safety and reliability estimation of automatic train protection and block system. *Quality and Reliability Engineering International*, 30(4), pp. 463–472.

214 Lee, P., Clark, A., Bushnell, L. and Poovendran, R., 2014. A passivity framework for modeling and mitigating wormhole attacks on networked control systems. *IEEE Transactions on Automatic Control*, 59(12), pp. 3224–3237.

215 Mitchell, R. and Chen, R., 2015. Modeling and analysis of attacks and counter defense mechanisms for cyber physical systems. *IEEE Transactions on Reliability*, 65(1), pp. 350–358.

216 Mitchell, R. and Chen, R., 2013. Effect of intrusion detection and response on reliability of cyber physical systems. *IEEE Transactions on Reliability*, 62(1), pp. 199–210.

217 Almalawi, A., Fahad, A., Tari, Z., Alamri, A., AlGhamdi, R. and Zomaya, A.Y., 2015. An efficient data-driven clustering technique to detect attacks in SCADA systems. *IEEE Transactions on Information Forensics and Security*, 11(5), pp. 893–906.

218 Shuang, Q., Zhang, M. and Yuan, Y., 2014. Node vulnerability of water distribution networks under cascading failures. *Reliability Engineering & System Safety*, 124, pp. 132–141.

219 Ryan, P.C., Stewart, M.G., Spencer, N. and Li, Y., 2014. Reliability assessment of power pole infrastructure incorporating deterioration and network maintenance. *Reliability Engineering & System Safety*, 132, pp. 261–273.

220 Srikantha, P. and Kundur, D., 2015. A DER attack-mitigation differential game for smart grid security analysis. *IEEE Transactions on Smart Grid*, 7(3), pp. 1476–1485.

221 Rocco, C.M., Ramirez-Marquez, J.E., Salazar, D.E. and Yajure, C., 2011. Assessing the vulnerability of a power system through a multiple objective contingency screening approach. *IEEE Transactions on Reliability*, 60(2), pp. 394–403.

222 Lyle, D., Chan, Y. and Head, E., 1999. Improving information-network performance: reliability versus invulnerability. *Iie Transactions*, 31(9), pp. 909–919.

223 Talarico, L., Reniers, G., Sörensen, K. and Springael, J., 2015. MISTRAL: A game-theoretical model to allocate security measures in a multi-modal chemical transportation network with adaptive adversaries. *Reliability Engineering & System Safety*, 138, pp. 105–114.

224 Hausken, K., 2008. Whether to attack a terrorist's resource stock today or tomorrow. *Games and Economic Behavior*, 64(2), pp. 548–564.

225 Hausken, K. and Zhuang, J., 2012. The timing and deterrence of terrorist attacks due to exogenous dynamics. *Journal of the Operational Research Society*, 63(6), pp. 726–735.

226 Zhuang, J., Bier, V.M. and Alagoz, O., 2010. Modeling secrecy and deception in a multiple-period attacker–defender signaling game. *European Journal of Operational Research*, 203(2), pp. 409–418.

227 Golalikhani, M. and Zhuang, J., 2011. Modeling arbitrary layers of continuous-level defenses in facing with strategic attackers. *Risk Analysis: An International Journal*, 31(4), pp. 533–547.

228 Golalikhani, M. and Zhuang, J., 2011. Modeling arbitrary layers of continuous-level defenses in facing with strategic attackers. *Risk Analysis: An International Journal*, 31(4), pp. 533–547.

229 Levitin, G. and Hausken, K., 2012. Individual versus overarching protection against strategic attacks. *Journal of the Operational Research Society*, 63(7), pp. 969–981.

230 Levitin, G. and Hausken, K., 2009. Parallel systems under two sequential attacks. *Reliability Engineering & System Safety*, 94(3), pp. 763–772.

231 Levitin, G. and Hausken, K., 2010. Resource distribution in multiple attacks against a single target. *Risk Analysis: An International Journal*, 30(8), pp. 1231–1239.

232 Jiang, L., Anantharam, V. and Walrand, J., 2010. How bad are selfish investments in network security?. *IEEE/ACM Transactions on Networking*, 19(2), pp. 549–560.

233 Bier, V.M., Gratz, E.R., Haphuriwat, N.J., Magua, W. and Wierzbicki, K.R., 2007. Methodology for identifying near-optimal interdiction strategies for a power transmission system. *Reliability Engineering & System Safety*, 92(9), pp. 1155–1161.

234 Zhai, Q., Ye, Z.S., Peng, R. and Wang, W., 2017. Defense and attack of performance-sharing common bus systems. *European Journal of Operational Research*, 256(3), pp. 962–975.

235 Van, P.D. and Bérenguer, C., 2012. Condition-based maintenance with imperfect preventive repairs for a deteriorating production system. *Quality and Reliability Engineering International*, 28(6), pp. 624–633.

236 Nagy, T. and Turanyi, T., 2012. Determination of the uncertainty domain of the Arrhenius parameters needed for the investigation of combustion kinetic models. *Reliability Engineering & System Safety*, 107, pp. 29–34.

237 Cedeño, E.B. and Arora, S., 2013. Cost impact of dynamically managing generation reserves. *International Journal of Electrical Power & Energy Systems*, 51, pp. 292–297.

238 Lau, A. and McSharry, P., 2010. Approaches for multi-step density forecasts with application to aggregated wind power. *The Annals of Applied Statistics*, pp. 1311–1341.

239 Yuan, W., Zhao, L. and Zeng, B., 2014. Optimal power grid protection through a defender–attacker–defender model. *Reliability Engineering & System Safety*, 121, pp. 83–89.

240 Torres, J.M., Brumbelow, K. and Guikema, S.D., 2009. Risk classification and uncertainty propagation for virtual water distribution systems. *Reliability Engineering & System Safety*, 94(8), pp. 1259–1273.

241 Wei, M., Jin, W. and Shen, L., 2012. A platoon dispersion model based on a truncated normal distribution of speed. *Journal of Applied Mathematics, 2012*.

242 Nan, C. and Sansavini, G., 2017. A quantitative method for assessing resilience of interdependent infrastructures. *Reliability Engineering & System Safety*, 157, pp. 35–53.

243 Tsai, H.Y. and Huang, Y.L., 2011. An analytic hierarchy process-based risk assessment method for wireless networks. *IEEE Transactions on Reliability*, 60(4), pp. 801–816.

244 Cho, J.H., Chen, R. and Feng, P.G., 2010. Effect of intrusion detection on reliability of mission-oriented mobile group systems in mobile ad hoc networks. *IEEE Transactions on Reliability*, 59(1), pp. 231–241.

245 Ntalampiras, S., 2014. Detection of integrity attacks in cyber-physical critical infrastructures using ensemble modeling. *IEEE Transactions on Industrial Informatics*, 11(1), pp. 104–111.

246 Konstantelos, I., Giannelos, S. and Strbac, G., 2016. Strategic valuation of smart grid technology options in distribution networks. *IEEE Transactions on Power Systems*, 32(2), pp. 1293–1303.

247 Pogaku, N., Prodanovic, M. and Green, T.C., 2007. Modeling, analysis and testing of autonomous operation of an inverter-based microgrid. *IEEE Transactions on power electronics*, 22(2), pp. 613–625.

248 Yu, K., Ai, Q., Wang, S., Ni, J. and Lv, T., 2015. Analysis and optimization of droop controller for microgrid system based on small-signal dynamic model. *IEEE Transactions on Smart Grid*, 7(2), pp. 695–705.

249 Jahromi, A.A., Kemmeugne, A., Kundur, D. and Haddadi, A., 2019. Cyber-physical attacks targeting communication-assisted protection schemes. *IEEE Transactions on Power Systems*, 35(1), pp. 440–450.

250 Khalili, M., Zhang, X., Cao, Y., Polycarpou, M.M. and Parisini, T., 2019. Distributed Fault-Tolerant Control of Multiagent Systems: An Adaptive Learning Approach. *IEEE Transactions on Neural Networks and Learning Systems*, 31(2), pp. 420–432.

251 Rana, M.M., Xiang, W. and Wang, E., 2018. Smart grid state estimation and stabilisation. *International Journal of Electrical Power & Energy Systems*, 102, pp. 152–159.

252 Todescato, M., Bof, N., Cavraro, G., Carli, R. and Schenato, L., 2020. Partition-based multi-agent optimization in the presence of lossy and asynchronous communication. *Automatica*, 111, p.108648.

253 Gallo, A.J., Turan, M.S., Boem, F., Parisini, T. and Ferrari-Trecate, G., 2020. A distributed cyber-attack detection scheme with application to DC microgrids. *IEEE Transactions on Automatic Control*, 65(9), pp. 3800–3815.

254 Mo, H. and Sansavini, G., 2017. Dynamic defense resource allocation for minimizing unsupplied demand in cyber-physical systems against uncertain attacks. *IEEE Transactions on Reliability*, 66(4), pp. 1253–1265.

255 Zhu, M. and Martínez, S., 2013. On distributed constrained formation control in operator–vehicle adversarial networks. *Automatica*, 49(12), pp. 3571–3582.

256 Che, L., Liu, X., Shuai, Z., Li, Z. and Wen, Y., 2018. Cyber cascades screening considering the impacts of false data injection attacks. *IEEE Transactions on Power Systems*, 33(6), pp. 6545–6556.

257 Patsakis, G., Rajan, D., Aravena, I., Rios, J. and Oren, S., 2018. Optimal black start allocation for power system restoration. *IEEE Transactions on Power Systems*, 33(6), pp. 6766–6776.

258 Rana, M.M., Li, L. and Su, S.W., 2017. Cyber attack protection and control of microgrids. *IEEE/CAA Journal of Automatica Sinica*, 5(2), pp. 602–609.

259 Smith, M.D. and Paté-Cornell, M.E., 2018. Cyber risk analysis for a smart grid: how smart is smart enough? a multiarmed bandit approach to cyber security investment. *IEEE Transactions on Engineering Management*, 65(3), pp. 434–447.

260 Zhang, X. and Papachristodoulou, A., 2015. A real-time control framework for smart power networks: Design methodology and stability. *Automatica*, 58, pp. 43–50.

261 Li, J.A., Dong, D., Wei, Z., Liu, Y., Pan, Y., Nori, F. and Zhang, X., 2020. Quantum reinforcement learning during human decision-making. *Nature Human Behaviour*, 4(3), pp. 294–307.

262 Littman, M.L., 2015. Reinforcement learning improves behaviour from evaluative feedback. *Nature*, 521(7553), pp. 445–451.

263 Sutton, R.S. and Barto, A.G., 2018. *Reinforcement learning: An introduction*. MIT press.

264 Fitouhi, M.C. and Nourelfath, M., 2012. Integrating noncyclical preventive maintenance scheduling and production planning for a single machine. *International Journal of Production Economics*, 136(2), pp. 344–351.

265 Lee, W.C., Wang, J.Y. and Lee, L.Y., 2015. A hybrid genetic algorithm for an identical parallel-machine problem with maintenance activity. *Journal of the Operational Research Society*, 66(11), pp. 1906–1918.

266 Schlünz, E.B. and Van Vuuren, J.H., 2013. An investigation into the effectiveness of simulated annealing as a solution approach for the generator maintenance scheduling problem. *International Journal of Electrical Power & Energy Systems*, 53, pp. 166–174.

267 Ahmad, M.A., Azuma, S.I. and Sugie, T., 2014. Performance analysis of model-free PID tuning of MIMO systems based on simultaneous perturbation stochastic approximation. *Expert Systems with Applications*, 41(14), pp. 6361–6370.

268 Chang, W.D. and Chen, C.Y., 2014. PID controller design for MIMO processes using improved particle swarm optimization. *Circuits, Systems, and Signal Processing*, 33(5), pp. 1473–1490.

269 Marseguerra, M., 2013. A MC-PSO approach to the failure probability evaluation of risky plant components: The maintenance design. *Reliability Engineering & System Safety*, 111, pp. 1–8.

270 Levitin, G., Xing, L. and Dai, Y., 2014. Mission cost and reliability of 1-out-of-$ N $ warm standby systems with imperfect switching mechanisms. *IEEE Transactions on Systems, Man, and Cybernetics: Systems*, 44(9), pp. 1262–1271.

271 Pan, I. and Das, S., 2013. Frequency domain design of fractional order PID controller for AVR system using chaotic multi-objective optimization. *International Journal of Electrical Power & Energy Systems*, 51, pp. 106–118.

272 Zhao, S.Z., Iruthayarajan, M.W., Baskar, S. and Suganthan, P.N., 2011. Multi-objective robust PID controller tuning using two lbests multi-objective particle swarm optimization. *Information Sciences*, 181(16), pp. 3323–3335.

273 Baghaee, H.R., Mirsalim, M., Gharehpetian, G.B. and Talebi, H.A., 2016. Reliability/cost-based multi-objective Pareto optimal design of stand-alone wind/PV/FC generation microgrid system. *Energy*, 115, pp. 1022–1041.

274 Baghaee, H.R., Mirsalim, M., Gharehpetian, G.B. and Kaviani, A.K., 2012. Security/cost-based optimal allocation of multi-type FACTS devices using multi-objective particle swarm optimization. *Simulation*, 88(8), pp. 999–1010.

275 Baghaee, H.R., Mirsalim, M. and Gharehpetian, G.B., 2017. Multi-objective optimal power management and sizing of a reliable wind/PV microgrid with hydrogen energy storage using MOPSO. *Journal of Intelligent & Fuzzy Systems*, 32(3), pp. 1753–1773.

276 Ye, Z., Revie, M. and Walls, L., 2014. A load sharing system reliability model with managed component degradation. *IEEE Transactions on Reliability*, 63(3), pp. 721–730.

277 Zhang, Z., Si, X., Hu, C. and Lei, Y., 2018. Degradation data analysis and remaining useful life estimation: A review on Wiener-process-based methods. *European Journal of Operational Research*, 271(3), pp. 775–796.

278 Al-Dabbagh, A.W. and Lu, L., 2010. Reliability modeling of networked control systems using dynamic flowgraph methodology. *Reliability Engineering & System Safety*, 95(11), pp. 1202–1209.

279 Liu, Y., He, X., Wang, Z. and Zhou, D., 2014. Optimal filtering for networked systems with stochastic sensor gain degradation. *Automatica*, 50(5), pp. 1521–1525.

280 Fan, J.H., Zhang, Y.M. and Zheng, Z.Q., 2012. Robust fault-tolerant control against time-varying actuator faults and saturation. *IET Control Theory & Applications*, 6(14), pp. 2198–2208.

281 Rajaram, M.L., Kougianos, E., Mohanty, S.P. and Choppali, U., 2016. Wireless sensor network simulation frameworks: A tutorial review: MATLAB/Simulink bests the rest. *IEEE Consumer Electronics Magazine*, 5(2), pp. 63–69.

282 Minero, P., Franceschetti, M., Dey, S. and Nair, G.N., 2009. Data rate theorem for stabilization over time-varying feedback channels. *IEEE Transactions on Automatic Control*, 54(2), pp. 243–255.

283 Cervin A, Henriksson D, Ohlin M. TrueTime 2.0-reference manual. <http://www .control.lth.se/truetime/>.

284 Gharavi, H. and Hu, B., 2015. Scalable synchrophasors communication network design and implementation for real-time distributed generation grid. *IEEE Transactions on Smart Grid*, 6(5), pp. 2539–2550.

285 Martin, K.E., Benmouyal, G., Adamiak, M.G., Begovic, M., Burnett, R.O., Carr, K.R., Cobb, A., Kusters, J.A., Horowitz, S.H., Jensen, G.R. and Michel, G.L., 1998. IEEE standard for synchrophasors for power systems. *IEEE Transactions on Power Delivery*, 13(1), pp. 73–77.

286 Kim, K.D. and Kumar, P.R., 2012. Real-time middleware for networked control systems and application to an unstable system. *IEEE Transactions on Control Systems Technology*, 21(5), pp. 1898–1906.

287 Fishman, G.S., 1986. A Monte Carlo sampling plan for estimating network reliability. *Operations Research*, 34(4), pp. 581–594.

288 Tsoutsanis, E., Meskin, N., Benammar, M. and Khorasani, K., 2016. A dynamic prognosis scheme for flexible operation of gas turbines. *Applied energy*, 164, pp. 686–701.

289 Pan, I. and Das, S., 2016. Fractional order fuzzy control of hybrid power system with renewable generation using chaotic PSO. *ISA transactions*, 62, pp. 19–29.

290 Bevrani, H., Feizi, M.R. and Ataee, S., 2015. Robust frequency control in an islanded microgrid: h∞ and μ-synthesis approaches. *IEEE transactions on smart grid*, 7(2), pp. 706–717.

291 Nagaraj, B. and Muruganath, N., 2010, October. A comparative study of PID controller tuning using GA, EP, PSO and ACO. In *2010 International Conference on Communication Control And Computing Technologies* (pp. 305–313). IEEE.

292 Kumar, A. and Gupta, R., 2013. Compare the results of Tuning of PID controller by using PSO and GA Technique for AVR system. *International Journal of Advanced Research in Computer Engineering & Technology (IJARCET)*, 2(6), pp. 2130–2138.

293 Mo, H. and Sansavini, G., 2018. Real-time coordination of distributed energy resources for frequency control in microgrids with unreliable communication. *International Journal of Electrical Power & Energy Systems*, 96, pp. 86–105.

294 Mo, H., Sansavini, G. and Xie, M., 2018. Performance-based maintenance of gas turbines for reliable control of degraded power systems. *Mechanical Systems and Signal Processing*, 103, pp. 398–412.

295 Baghaee, H.R., Mirsalim, M., Gharehpetian, G.B. and Talebi, H.A., 2017. Application of RBF neural networks and unscented transformation in probabilistic power-flow of microgrids including correlated wind/PV units and plug-in hybrid electric vehicles. *Simulation Modelling Practice and Theory*, 72, pp. 51–68.

296 Kaviani, A.K., Baghaee, H.R. and Riahy, G.H., 2009. Optimal sizing of a stand-alone wind/photovoltaic generation unit using particle swarm optimization. *Simulation*, 85(2), pp. 89–99.

297 Baghaee, H.R., Mirsalim, M., Gharehpetian, G.B. and Talebi, H.A., 2017. Fuzzy unscented transform for uncertainty quantification of correlated wind/PV microgrids: possibilistic–probabilistic power flow based on RBFNNs. *IET Renewable Power Generation*, 11(6), pp. 867–877.

298 Galbusera, L., Theodoridis, G. and Giannopoulos, G., 2014. Intelligent energy systems: Introducing power–ICT interdependency in modeling and control design. *IEEE Transactions on Industrial Electronics*, 62(4), pp. 2468–2477.

299 Sahu, R.K., Panda, S. and Sekhar, G.C., 2015. A novel hybrid PSO-PS optimized fuzzy PI controller for AGC in multi area interconnected power systems. *International Journal of Electrical Power & Energy Systems*, 64, pp. 880–893.

300 Baghaee, H.R., Mirsalim, M., Gharehpetian, G.B., Talebi, H.A. and Niknam-Kumle, A., 2017. A hybrid ANFIS/ABC-based online selective harmonic elimination switching pattern for cascaded multi-level inverters of microgrids. *IEEE Trans. Ind. Electron*, 99, pp. 1–10.

301 Chaudhuri, N.R., Ray, S., Majumder, R. and Chaudhuri, B., 2009. A new approach to continuous latency compensation with adaptive phasor power oscillation damping controller (POD). *IEEE Transactions on Power Systems*, 25(2), pp. 939–946.

302 Allen, A., Santoso, S. and Muljadi, E., 2013. *Algorithm for screening phasor measurement unit data for power system events and categories and common characteristics for events seen in phasor measurement unit relative phase-angle differences and frequency signals* (No. NREL/TP-5500-58611). National Renewable Energy Lab.(NREL), Golden, CO (United States).

303 Power System Relaying Committee, 2011. C37. 118.1-2011: Ieee standard for synchrophasor measurements for power systems. *IEEE Standard Association*.

304 Ray, P.K., Mohanty, S.R. and Kishor, N., 2011. Proportional–integral controller based small-signal analysis of hybrid distributed generation systems. *Energy Conversion and Management*, 52(4), pp. 1943–1954.

305 Sharma, G., Nasiruddin, I., Niazi, K.R. and Bansal, R.C., 2016. Optimal AGC of a multi-area power system with parallel AC/DC tie lines using output vector feedback control strategy. *International Journal of Electrical Power & Energy Systems*, 81, pp. 22–31.

306 Liang, Y., Chen, T. and Pan, Q., 2010. Optimal linear state estimator with multiple packet dropouts. *IEEE Transactions on Automatic Control*, 55(6), pp. 1428–1433.

307 Singh, V.P., Kishor, N. and Samuel, P., 2016. Load frequency control with communication topology changes in smart grid. *IEEE Transactions on Industrial Informatics*, 12(5), pp. 1943–1952.

308 ENTSO-e, O.H., 2009. P1-Policy 1: Load-Frequency Control and Performance.

309 Mo, H. and Sansavini, G., 2021. Hidden Markov model-based smith predictor for the mitigation of the impact of communication delays in wide-area power systems. *Applied Mathematical Modelling*, 89, pp. 19–48.

310 Fovino, I.N., Coletta, A., Carcano, A. and Masera, M., 2011. Critical state-based filtering system for securing SCADA network protocols. *IEEE Transactions on industrial electronics*, 59(10), pp. 3943–3950.

311 Mo, H., Wang, W., Xie, M. and Xiong, J., 2015. Modeling and analysis of the reliability of digital networked control systems considering networked degradations. *IEEE Transactions on Automation Science and Engineering*, 14(3), pp. 1491–1503.

312 Zhao, J. and Wang, J., 2013. Control-oriented multi-phase combustion model for biodiesel fueled engines. *Applied energy*, 108, pp. 92–99.

313 Yue, D. and Han, Q.L., 2005. Delayed feedback control of uncertain systems with time-varying input delay. *Automatica*, 41(2), pp. 233–240.

314 Darabi, Z. and Ferdowsi, M., 2012. Impact of plug-in hybrid electric vehicles on electricity demand profile. In *Smart Power Grids 2011* (pp. 319–349). Springer, Berlin, Heidelberg.

315 Yue, D., Tian, E., Zhang, Y. and Peng, C., 2008. Delay-distribution-dependent stability and stabilization of T–S fuzzy systems with probabilistic interval delay. *IEEE Transactions on Systems, Man, and Cybernetics, Part B (Cybernetics)*, 39(2), pp. 503–516.

316 Kim, D.S., Lee, Y.S., Kwon, W.H. and Park, H.S., 2003. Maximum allowable delay bounds of networked control systems. *Control Engineering Practice*, 11(11), pp. 1301–1313.

317 Li, L. and Zhong, L., 2014. Generalised nonlinear l 2–l∞ filtering of discrete-time Markov jump descriptor systems. *International Journal of Control*, 87(3), pp. 653–664.

318 Zhang, B. and Zheng, W.X., 2012. H∞ filter design for nonlinear networked control systems with uncertain packet-loss probability. *Signal Processing*, 92(6), pp. 1499–1507.

319 Gonçalves, A.P., Fioravanti, A.R. and Geromel, J.C., 2011. Filtering of discrete-time Markov jump linear systems with uncertain transition probabilities. *International Journal of Robust and Nonlinear Control*, 21(6), pp. 613–624.

320 Henrion, D. and Lasserre, J.B., 2011. Inner approximations for polynomial matrix inequalities and robust stability regions. *IEEE Transactions on Automatic Control*, 57(6), pp. 1456–1467.

321 Aalen, O., 1978. Nonparametric estimation of partial transition probabilities in multiple decrement models. *The Annals of Statistics, pp.*534–545.

322 Mena, R., Hennebel, M., Li, Y.F., Ruiz, C. and Zio, E., 2014. A risk-based simulation and multi-objective optimization framework for the integration of distributed renewable generation and storage. *Renewable and Sustainable Energy Reviews*, 37, pp. 778–793.

323 Borges, C.L.T., 2012. An overview of reliability models and methods for distribution systems with renewable energy distributed generation. *Renewable and sustainable energy reviews*, 16(6), pp. 4008–4015.

324 Li, Y.F. and Zio, E., 2012. A multi-state model for the reliability assessment of a distributed generation system via universal generating function. *Reliability Engineering & System Safety*, 106, pp. 28–36.

325 Elia Grid, Belgium, from January 1, 2015 to September 22, 2015. Available at: http://www.elia.be/en/grid-data/data-download.

326 Shafiee, S., Fotuhi-Firuzabad, M. and Rastegar, M., 2013. Investigating the impacts of plug-in hybrid electric vehicles on power distribution systems. *IEEE Transactions on Smart Grid*, 4(3), pp. 1351–1360.

327 Samaras, C. and Meisterling, K., 2008. Life cycle assessment of greenhouse gas emissions from plug-in hybrid vehicles: implications for policy.

328 Meng, J., Mu, Y., Jia, H., Wu, J., Yu, X. and Qu, B., 2016. Dynamic frequency response from electric vehicles considering travelling behavior in the Great Britain power system. *Applied energy*, 162, pp. 966–979.

329 Mena, R., Hennebel, M., Li, Y.F. and Zio, E., 2014. Self-adaptable hierarchical clustering analysis and differential evolution for optimal integration of renewable distributed generation. *Applied Energy*, 133, pp. 388–402.

330 Díaz-González, F., Sumper, A., Gomis-Bellmunt, O. and Villafáfila-Robles, R., 2012. A review of energy storage technologies for wind power applications. *Renewable and sustainable energy reviews*, 16(4), pp. 2154–2171.

331 Luo, X., Wang, J., Dooner, M. and Clarke, J., 2015. Overview of current development in electrical energy storage technologies and the application potential in power system operation. *Applied energy*, 137, pp. 511–536.

332 Li, Z., Guo, Q., Sun, H. and Wang, J., 2015. Storage-like devices in load leveling: Complementarity constraints and a new and exact relaxation method. *Applied Energy*, 151, pp. 13–22.

333 Al-Wakeel, A., Wu, J. and Jenkins, N., 2017. K-means based load estimation of domestic smart meter measurements. *Applied energy*, 194, pp. 333–342.

334 Orwig, K., 2010. Examining strong winds from a time-varying perspective (Doctoral dissertation).

335 Montoya, F.G., García-Cruz, A., Montoya, M.G. and Manzano-Agugliaro, F., 2016. Power quality techniques research worldwide: A review. *Renewable and Sustainable Energy Reviews*, 54, pp. 846–856.

336 Widén, J., Carpman, N., Castellucci, V., Lingfors, D., Olauson, J., Remouit, F., Bergkvist, M., Grabbe, M. and Waters, R., 2015. Variability assessment and forecasting of renewables: A review for solar, wind, wave and tidal resources. *Renewable and Sustainable Energy Reviews*, 44, pp. 356–375.

337 Arce, M.E., Saavedra, Á., Míguez, J.L. and Granada, E., 2015. The use of grey-based methods in multi-criteria decision analysis for the evaluation of sustainable energy systems: A review. *Renewable and Sustainable Energy Reviews*, 47, pp. 924–932.

338 Foley, A.M., Leahy, P.G., Marvuglia, A. and McKeogh, E.J., 2012. Current methods and advances in forecasting of wind power generation. *Renewable Energy*, 37(1), pp. 1–8.

339 Julong, D., 1989. Introduction to grey system theory. *The Journal of grey system*, 1(1), pp. 1–24.

340 Tien, T.L., 2009. A new grey prediction model FGM (1, 1). *Mathematical and Computer Modelling*, 49(7-8), pp. 1416–1426.

341 Bahrami, S., Hooshmand, R.A. and Parastegari, M., 2014. Short term electric load forecasting by wavelet transform and grey model improved by PSO (particle swarm optimization) algorithm. *Energy*, 72, pp. 434–442.

342 Zhao, J., Wang, J. and Su, Z., 2014. Power generation and renewable potential in China. *Renewable and Sustainable Energy Reviews*, 40, pp. 727–740.

343 Hsu, C.C. and Chen, C.Y., 2003. Applications of improved grey prediction model for power demand forecasting. *Energy Conversion and management*, 44(14), pp. 2241–2249.

344 Martins, V.F. and Borges, C.L., 2011. Active distribution network integrated planning incorporating distributed generation and load response uncertainties. *IEEE Transactions on power systems*, 26(4), pp. 2164–2172.

345 Zio, E., 2013. System reliability and risk analysis. In *The Monte Carlo simulation method for system reliability and risk analysis* (pp. 7–17). Springer, London.

346 Ren, H., Zhou, W., Nakagami, K.I., Gao, W. and Wu, Q., 2010. Multi-objective optimization for the operation of distributed energy systems considering economic and environmental aspects. *Applied Energy*, 87(12), pp. 3642–3651.

347 Son, G.T., Lee, H.J., Nam, T.S., Chung, Y.H., Lee, U.H., Baek, S.T., Hur, K. and Park, J.W., 2012. Design and control of a modular multilevel HVDC converter with redundant power modules for noninterruptible energy transfer. *IEEE Transactions on Power Delivery*, 27(3), pp. 1611–1619.

348 Hu, P., Jiang, D., Zhou, Y., Liang, Y., Guo, J. and Lin, Z., 2013. Energy-balancing control strategy for modular multilevel converters under submodule fault conditions. *IEEE Transactions on Power Electronics*, 29(9), pp. 5021–5030.

349 IEEE Power and Energy Society. Distribution test feeders <http://ewh.ieee.org/soc/pes/dsacom/testfeeders/index.html>.

350 Kersting, W.H., 1991. Radial distribution test feeders. *IEEE Transactions on Power Systems*, 6(3), pp. 975–985.

351 Chen, C.I. and Chen, Y.C., 2013. Comparative study of harmonic and interharmonic estimation methods for stationary and time-varying signals. *IEEE Transactions on Industrial Electronics*, 61(1), pp. 397–404.

352 Wang, D. and Peter, W.T., 2015. Prognostics of slurry pumps based on a moving-average wear degradation index and a general sequential Monte Carlo method. *Mechanical systems and signal processing*, 56, pp. 213–229.

353 Jin, T. and Mechehoul, M., 2010. Minimize production loss in device testing via condition-based equipment maintenance. *IEEE transactions on automation science and engineering*, 7(4), pp. 958–963.

354 Wu, F., Wang, T. and Lee, J., 2010. An online adaptive condition-based maintenance method for mechanical systems. *Mechanical Systems and Signal Processing*, 24(8), pp. 2985–2995.

355 Shafiee, M., Finkelstein, M. and Bérenguer, C., 2015. An opportunistic condition-based maintenance policy for offshore wind turbine blades subjected to degradation and environmental shocks. *Reliability Engineering & System Safety*, 142, pp. 463–471.

356 Zhang, H., Dai, H., Beer, M. and Wang, W., 2013. Structural reliability analysis on the basis of small samples: an interval quasi-Monte Carlo method. *Mechanical Systems and Signal Processing*, 37(1-2), pp. 137–151.

357 Lake Cogen Ltd, USA, 60.5 MW, [online] Available: <http://www.energyjustice.net/map/displayfacility-68186.htm>.

358 Fitiwi, D.Z., Olmos, L., Rivier, M., De Cuadra, F. and Pérez-Arriaga, I.J., 2016. Finding a representative network losses model for large-scale transmission expansion planning with renewable energy sources. *Energy*, 101, pp. 343–358.

359 IEEE 13 nodes test feeder. Available at: http://sites.ieee.org.wwwproxy1.library.unsw.edu.au/pes-testfeeders/.

360 Staffell, I. and Green, R., 2014. How does wind farm performance decline with age?. *Renewable energy*, 66, pp. 775–786.

361 Alvarez-Alvarado, M.S. and Jayaweera, D., 2018, June. Aging reliability model for generation adequacy. In *2018 IEEE International Conference on Probabilistic Methods Applied to Power Systems (PMAPS)* (pp. 1–6). IEEE.

362 AgriMet Weather Database. Available at https://www.usbr.gov/pn/agrimet/wxdata.html.

363 FERC Form 1 Database. Available at: https://www.ferc.gov/docs-filing/forms/form-1/viewer-instruct.asp.

364 Everitt, B. and Skrondal, A., 2002. *The Cambridge dictionary of statistics* (Vol. 106). Cambridge: Cambridge University Press.

365 Kumar, N., Besuner, P., Lefton, S., Agan, D. and Hilleman, D., 2012. *Power plant cycling costs* (No. NREL/SR-5500-55433). National Renewable Energy Lab.(NREL), Golden, CO (United States).

366 Mo, H. and Sansavini, G., 2019. Impact of aging and performance degradation on the operational costs of distributed generation systems. *Renewable energy*, 143, pp. 426–439.

367 Zhao, W., Tao, T., Zio, E. and Wang, W., 2016. A novel hybrid method of parameters tuning in support vector regression for reliability prediction: particle swarm optimization combined with analytical selection. *IEEE Transactions on Reliability*, 65(3), pp. 1393–1405.

368 Hausken, K. and Levitin, G., 2012. Review of systems defense and attack models. *International Journal of Performability Engineering*, 8(4), pp. 355–366.

369 Xu, X., Li, Z. and Chen, N., 2015. A hierarchical model for lithium-ion battery degradation prediction. *IEEE Transactions on Reliability*, 65(1), pp. 310–325.

370 Li, G.D., Masuda, S., Yamaguchi, D. and Nagai, M., 2010. A new reliability prediction model in manufacturing systems. *IEEE Transactions on Reliability*, 59(1), pp. 170–177.

371 Pillitteri, V.Y. and Brewer, T.L., 2014. *Guidelines for smart grid cybersecurity* (No. NIST Interagency/Internal Report (NISTIR)-7628 Rev 1).

372 Amin, S., Schwartz, G.A. and Sastry, S.S., 2013. Security of interdependent and identical networked control systems. *Automatica*, 49(1), pp. 186–192.

373 Lam, C.T. and Lehoczky, J.P., 1991. Superposition of renewal processes. *Advances in applied probability*, pp. 64–85.

374 Freund, Y. and Schapire, R.E., 1999. Adaptive game playing using multiplicative weights. *Games and Economic Behavior*, 29(1-2), pp. 79–103.

375 Bayesian adversarial multi-node bandit for optimal smart grid protection against cyber-attacks.

376 Osband, I., Russo, D. and Van Roy, B., 2013. (More) efficient reinforcement learning via posterior sampling. In *Advances in Neural Information Processing Systems* (pp. 3003-3011).

377 Besbes, O., Gur, Y. and Zeevi, A., 2019. Optimal exploration–exploitation in a multi-armed bandit problem with non-stationary rewards. *Stochastic Systems*, 9(4), pp. 319–337.

378 Yang, G., 2009. Reliability demonstration through degradation bogey testing. *IEEE Transactions on Reliability*, 58(4), pp. 604–610.

379 Yang, Y.J., Peng, W., Meng, D., Zhu, S.P. and Huang, H.Z., 2014. Reliability analysis of direct drive electrohydraulic servo valves based on a wear degradation process and individual differences. *Proceedings of the Institution of Mechanical Engineers, Part O: Journal of Risk and Reliability*, 228(6), pp. 621–630.

380 Xu, D., Feng, Z., Sui, S. and Lin, Y.H., 2019. Reliability Assessment of Electrohydraulic Actuation Control System Subject to Multisources Degradation Processes. *IEEE/ASME Transactions on Mechatronics*, 24(6), pp. 2594–2605.

381 Lawless, J. and Crowder, M., 2004. Covariates and random effects in a gamma process model with application to degradation and failure. *Lifetime data analysis*, 10(3), pp. 213–227.

382 Ye, Z.S. and Chen, N., 2014. The inverse Gaussian process as a degradation model. *Technometrics*, 56(3), pp. 302–311.

383 Reliability Analysis of Aging Control System via Stability Margins

384 Dong, Q. and Cui, L., 2019. First hitting time distributions for Brownian motion and regions with piecewise linear boundaries. *Methodology and Computing in Applied Probability*, 21(1), pp. 1–23.

385 Si, X.S., Ren, Z., Hu, X., Hu, C.H. and Shi, Q., 202. A novel degradation modelling and prognostic framework for closed-loop systems with degrading actuator. *IEEE Transactions on Industrial Electronics*, 67(11), pp. 9635–9647.

386 Normey-Rico, J.E., 2007. *Control of dead-time processes.* Springer Science & Business Media.

387 Kao, C.Y. and Lincoln, B., 2004. Simple stability criteria for systems with time-varying delays. *Automatica*, 40(8), pp. 1429–1434.

388 Nash, J., 1953. Two-person cooperative games. *Econometrica: Journal of the Econometric Society*, pp. 128–140.

389 Attiah, A., Chatterjee, M. and Zou, C.C., 2018, May. A game theoretic approach to model cyber attack and defense strategies. In *2018 IEEE International Conference on Communications (ICC)* (pp. 1–7). IEEE.

390 Do, C.T., Tran, N.H., Hong, C., Kamhoua, C.A., Kwiat, K.A., Blasch, E., Ren, S., Pissinou, N. and Iyengar, S.S., 2017. Game theory for cyber security and privacy. *ACM Computing Surveys (CSUR)*, 50(2), pp. 1–37.

391 Game Attack-Defense Graph Approach for Modelling and Analysis of Cyber-Attacks and Defenses in Local Metering System

Index

Cyber-Physical Distributed Systems: Modeling, Reliability Analysis and Applications, First Edition.
Huadong Mo, Giovanni Sansavini and Min Xie.
© 2021 John Wiley & Sons Ltd. Published 2021 by John Wiley & Sons Ltd.